THE
APPLE
BOOK

THE APPLE BOOK

ROSIE SANDERS

FRANCES LINCOLN LIMITED

PUBLISHERS

For Harry

The silver apples of the moon,
The golden apples of the sun.

From 'The Song of the Wandering Aengus' by W.B Yeats

Frances Lincoln Limited
4 Torriano Mews
Torriano Avenue
London NW5 2RZ
www.franceslincoln.com

First Frances Lincoln edition 2010

Royalties from the sale of this book go towards the charitable
work of the RHS, promoting horticulture and helping gardeners.
www.rhs.org.uk

A catalogue record for this book is available from the
British Library.

978-0-7112-3141-2

Printed and bound in Singapore

9 8 7 6 5 4 3 2 1

HALF TITLE PAGE *Golden Noble*
TITLE PAGE *Bess Pool*

CONTENTS

INTRODUCTION

I love painting apples. I never get bored with it. They are so individual in their character and there is something about their solidity and shape, that lovely roundness that creates exciting edges and reflections. Then there are the colours, the patterns and texture, the milky white transparent layer that sometimes overlays the colour, the golden russet that coats some like a garment or breaks and scatters revealing patches of glowing colours beneath. Then there is the shine on some of them and the challenge of painting the illusion of all those highlights and reflections, casting a round object on to a flat piece of paper: Like the infinite stars, no one apple is the same as another.

And the orchard, what could be more enticing than a traditional orchard with the diversity of wildlife it supports? The colour of the apples vibrating in the sunshine against the green of the leaves is a feast for the eyes and the sounds of the birds and insects, the occasional thud as an apple falls on to the grass, music to the ears. Peaceful and secure, they speak of quieter times when life was not so hurried and draw you back to the earth and its beneficence. There is an expectation too in taking an apple in your fingers, giving it a little twist and feeling it drop into the palm of your hand, smooth and warm, a portable meal so full of flavour and goodness. No wonder Eve was so tempted. There is such a rich cultural history associated with apples, so many stories, customs and beliefs, it seems a shame that most of us can't name more than about ten varieties. A shame too that we generally don't get the opportunity to experience the quality and variety of flavours.

It is uncertain when apples were first cultivated in Britain. but possibly varieties of apple were introduced here during the Roman occupation. There is no doubt that apples grew in England before then as both Celtic and Cornish dialects have words for apple, but it is almost certainly the indigenous crab, *Malus sylvestris,* to which they referred. The Romans had discovered the art of grafting and budding and cultivated new varieties from the East. These probably spread through Europe with the occupying forces and ultimately, it is believed, came to England.

It was not until medieval times that fruit culture became firmly established. The monasteries introduced new apples from France both for the production of cider and for dessert. By the end of the thirteenth century, several varieties were established and among them were the first properly recorded apples. The Old English Pearmain, recorded in 1204, was the main dessert apple in England until well into the eighteenth century. Sadly, today this apple has almost disappeared.

Its culinary partner was the Costard, recorded as being sold in the markets of Oxford in 1296. Although the most important culinary apple until the end of the seventeenth century, the only souvenir that remains of it is the word 'costermonger'. The period of prosperity during the thirteenth century was cut short in the early 1300s by succesive droughts, the Black Death, and the Wars of the Roses..., so that eventually the general demise of the rural areas led to a reduction in fruit growing. Apples were then imported mainly by those who could afford them and this continued until the sixteenth century and the reign of Henry VIII.

Henry VIII instructed his fruiterer Richard Harris to bring over the best varieties of fruit from France. Harris was responsible for importing numerous grafts and this was an important turning point in fruit growing in England. His orchard at Teynham in Kent became the foundation for other orchards, representing a basis for the propogation of new cultivars. By the end of the 1500s market gardens were established in Kent and Surrey.

Apple culture continued rather haphazardly until the start of the 1800s when Thomas Andrew Knight pioneered a scientific approach to apple breeding. He began a series of experiments raising seedlings from two selected parents, believing that each apple cultivar had a definite lifespan beyond which neither the original tree nor any graft taken from it would live. Although mistaken, his work suggested the possibilities of selective apple breeding.

In the propagation of apples we owe a great deal to the inspired and gifted amateur. Many apples came into being at the gentle hands of a gardener, or were conceived by the chance planting of a pip in some cottage garden. Bramley's Seedling, famous throughout the world, was raised in a garden in Nottinghamshire. The Blenheim Orange was raised by Mr. Kempster, a labourer from Oxfordshire and more recently Discovery was raised in Essex by Mr. Dummer.

During the early part of the 19th Century, fruit growing was expanding, and with the new technology of growing under glass, interest in growing exotic fruit became a passion. In Victorian times however, interest returned to apples, perhaps because with careful selection and planting, apples could provide a continuing source of variety throughout the year. Apples became a prized dessert fruit and were discussed and written about passionately by discerning Victorians. Apples were grown in ever more diverse ways such as cordons and espaliers, providing screens and borders for walks around the kitchen gardens. Horticulture became the hobby of many a retired businessman and the most famous apple of all, the Cox's Orange Pippin, came about in this way. Richard Cox, a successful brewer, retired in 1820 and together with three gardeners devoted his remaining years to horticulture and thus became immortalized both through Cox's Orange Pippin and Cox's Pomona.

Amateurs and nurserymen alike are remembered through their apples. None can be more famous than the Laxton family. Thomas Laxton worked hard to improve the quality of pears and strawberries and his sons extended his work to apples. In 1854 the British Pomological Society was set up to promote fruit culture and classify apple cultivars in Britain, Europe and America. Subsequently, interest in apple growing was confirmed at The Great Apple and Pear Show staged by the Royal Horticultural Society (RHS) in 1883, when the largest collection of apples ever assembled was brought from gardens and orchards throughout the land. Since its inception in 1804 the RHS has been able to call upon the advice of the most eminent nurserymen of the day, and today it identifies the hundreds of apples that are sent to it every year.

After World War 1, it was realized that English orchards grew a mixture of cultivars and that apple breeding was rather indiscriminate, making commercial growing unnecessarily difficult. In October 1922, after discussions with the Ministry of Agriculture, the RHS set in motion the establishment of the first ever commercial fruit trials. The purpose was to carry out impartial trials of the value of cultivars and to inform growers of the new introductions and the susceptibility of cultivars to disease. National fruit trials were establised at Wisley. After Word War 11 they became part of the National Agricultural Advisory Service and moved to Brogdale Farm near Faversham in Kent. The transfer was completed in 1960.

Today the majority of top cultivars come from the proffessional plant breeders such as East Malling Research Station, the John Innes Institute, Long Ashton Research Station, the Scottish Horticultural Research Institute and research stations abroad. The whole business has been put on a very professional footing but, although the results are excellent, the romance has gone. Apples are no longer named after characters such as the much loved Reverend W Wilks, William Crump, or Lady Sudeley, whose dress it was that inspired her husband to name after her the apple his gardener had raised, or Bess Pool, named after the girl who found a new apple while wandering through a wood.

But there are encouraging signs that we are seeing a revival of the interest in apples. On the 21st October 1990, Common Ground organised the first Apple Day at Covent Garden in London, an event that has initiated an ever increasing annual programme of apple events around Britain. They are an internationally recognised charity who, among other things, offer help in setting up local community orchards to provide reservoirs for local apple varieties, help restore old orchards to try and halt the terrible decline (67% in 27 years) of Britain's commercial orchards. There does seem to be a healthy interest in all aspects of home produce however, whether it is in allotments, community orchards, or farmer's markets. I hope that this book may play a part in this renaissance.

THE IDENTIFICATION OF APPLES

SEASON

The apples are listed according to season, starting with the earliest. The season given is that during which the fruit can be eaten at its best. The early cultivars will not keep and cannot be stored, the mid-season cultivars will only keep for a short period, but the late cultivars only mature some time after they are picked and need to be stored before they reach their full flavour.

Ripening of the fruit is subject to regional and seasonal variation, so precise dates for picking are not possible. The best test for readiness is to lift the fruit gently and give it a slight twist. If it parts easily from the spur, leaving the stalk intact, it is ready to be picked.

FRUIT CHARACTERISTICS

Size

The size was taken from the fruit of old mature trees. Fruit taken from young trees are almost invariably larger than that quoted, especially if they are on dwarfing rootstocks. An average size has been estimated and given as diameter first and height second. The sizes are grouped into seven categories.

Fruit size will vary according to the climate, soil and size of the crop and the sizes given are intended only as a general guide

King Fruit

This fruit lies at the centre of the truss. Being the central flower it is the first to open and the fruit is often larger than the others. It sometimes has a fleshy protuberance at the stalk end and often the stalk is very short and thick. When picking the fruit for identification it is better, if possible, to avoid the king fruit as it may be misleading.

Shape

The shapes have been defined under eight headings. The shape of an apple can be a fairly regular guide, being only slightly influenced by climate, soil or crop size. Fruits from the northern latitudes are sometimes slightly more conical. When examining the shape of an apple, it should be held with the stalk (base) downwards and the eye (apex) uppermost.

Ribs Are apparent on some cultivars. Up to five ribs can run between base and apex.
Five crowned Sometimes the ribs become a pronounced 'crown' at the apex.

Symmetrical When the sides are equally developed and the fruit is symmetrical.
Lop-sided When the sides have developed unequally.
Regular When a horizontal section appears to be nearly circular.
Irregular When a horizontal section is angular, elliptical or irregular.
Waisted When the curve is concave towards the apex, for example Laxton's Early Crimson.

Skin

The descripton is intended to define a good example of the fruit showing the characteristics normally found on that particular cultivar. The examples chosen are fruits that have been well exposed to the sun and not those which have been shaded, as these will tend to be greener and lacking in the characteristic colours. Inevitably variances will be found according to the amount of sun and the type of soil. The examples illustrated show ripe fruits as they appear on the tree. Some fruits become yellower when stored and some become more greasy and shiny, and if so this is stated in the text.

Flush This is an area of almost unbroken colour which can appear as a small patch or almost wholly cover the fruit.
Mottling When the overlying colour is broken showing the skin beneath.
Stripes The fruit can be striped with varying shades of red. These stripes can be well defined long unbroken stripes, such as appear in Jupiter, or short broken stripes or splashes, as in Queen.
Scarf Skin This term is used to describe a thin whitish layer of skin giving a rather milky appearance to part of the fruit, usually around the base, as in Grenadier and Jupiter. Not to be confused with bloom which rubs off.
Hammering Sometimes the surface is slightly pitted, rather like beaten copper, and feels slightly uneven. This is a characteristic sometimes observed in cultivars such as Jonathan and Annie Elizabeth.
Hair Line Some cultivars occasionally have fruits that bear a narrow, sometimes russetted, hair line that runs between base and apex. It is not a malformation but a characteristic of that cultivar. It is found, for example, in Keswick Codlin.
Russet This is a brown, somewhat cork-like layer that forms over the skin of the apple. Russet can appear in small or large patches, small scuffs, or resembling scattered dust. It can almost entirely cover the surface of the fruit and in this case the cultivar is known as a Russet apple, for example Egremont Russet.

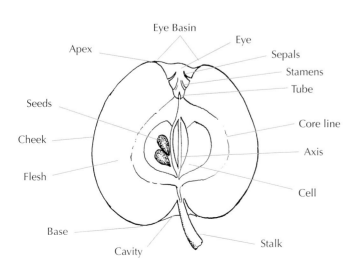

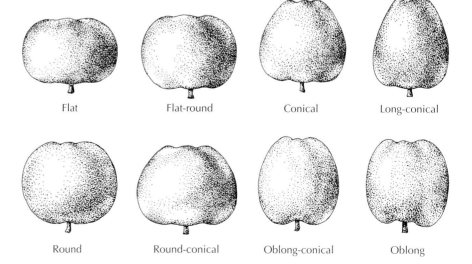

The amount and location of russet on the fruit is important for identification; for example, on Rosemary Russet it is mainly concentrated around the apex

Reinette A group of apples with colour and with areas of russet but not in sufficient quantity for the apple to be classified as a Russet apple. The term was originally intended to denote a fruit of quality, as indicated by the finest Reinette of all, the Cox's Orange Pippin.

Netting A network of russet can sometimes extend over part of the surface of the fruit, as in Lord Hindlip.

Lenticels These are pores that are irregularly distributed over the surface of the fruit through which gasses interchange between the interior and exterior surfaces. They are an important feature of identification. They are usually roundish but can be angular or star-shaped, as is sometimes found in Sunset and Charles Ross.

Areolar In some cultivars, such as Granny Smith, a halo of colour can surround the lenticel.

Texture A fruit can be said to be smooth as in Lane's Prince Albert, or rough as in Cornish Gilliflower. It can be dry as in Grenadier, or greasy as in Bramley's Seedling. This grease is a wax produced naturally by the fruit, as opposed to the protective wax used commercially, and often increases after the fruit has been picked and stored.

Bloom In some cultivars, such as McIntosh Red and Red Delicious, the surface of the skin is covered with a fine whitish bloom. This can easily be rubbed off so it is more apparent when the fruit is still on the tree.

Stalk

The identifying points about the stalk are length and thickness. These can vary between fruits of any cultivar but there is a general expectation in most cases. Sometimes however, definite variations occur, as in Ellison's Orange, where the stalk can either be fairly long and thin or short and stoutish. Some cultivars have fruits with a fleshy bump on the stalk which is a characteristic and not a malformation, for example Newton Wonder. The stalk of the king fruit (see sub-heading 'Size') can be fleshy or swollen.

Cavity

The depression round the stalk is known as the cavity. The depth and width and the presence or absence of russet is important.

Lipped Some fruits within a cultivar can develop a lip on one side of the cavity which often results in the stalk being set at an angle. Good examples are Jupiter and Charles Ross.

Eye

This is the only part of the flower remaining after the fruit has been formed. After the petals drop the fruit develops and when mature the five sepals take up one of six distinct forms.

Erect convergent in which the sepals are standing upright with their margins touching and their points meeting together at a common point.

Erect in which the sepals are standing upright but not meeting together.

Connivent in which the sepals are standing upright with their points meeting together but overlapping each other.

Flat convergent in which the sepals lie flat with their points facing inwards and their margins touching.

Divergent in which the sepals are quite reflexed and fall back on to the basin. The eye can be open, partly open or closed.

Reflexed tips The tips of the sepals can be bent outwards away from the eye.

Basin

The depression in which the eye is situated is called the basin Ribs may be visible within the basin and sometimes the skin is puckered around the eye.

Beading In some cultivars, such as Merton Worcester, up to five small round fleshy knobs are present in the basin round the eye.

Downy Varying amounts of white downy hairs can be present on the sepals.

INTERNAL CHARACTERISTICS

Robert Hogg in *The Fruit Manual* based his classification on the internal characteristics of the fruit and I have followed his principles of description: the shape of the tube, the position of the stamens within the tube, the shape of the carpels or seed cells and their relationship to the axis. All these internal characters, with the exception of the relationship of the cells to the axis, can be seen in the drawn cross section of each cultivar.

Tube The cavity just beneath the eye, (a) cone-shaped, (b) funnel-shaped.

Stamens
  (a) marginal–when they are situated at the outer margin of the tube,
  (b) median–when they are situated approximately in the middle of the tube,
  (c) basal–when they lie near the base of the tube.

Core line A line surrounding the core is usually clearly visible and its point of attachment to the tube can vary:
  (a) Basal–meeting (when the lines either side of the tube meet and touch at the base of the tube).
  (b) Basal–clasping (when the lines either side of the tube join the base of the tube but do not touch).
  (c) Median–(when the lines either side of the tube join the tube in the middle).
  (d) Marginal–(when the lines either side of the tube join the tube at the point nearest the eye).

Core The position of the core can vary. It can be close to the stalk (sessile), at the centre of the fruit (median), or far from the stalk (distant).

Cells If the seed cell is split through the middle longitudinally it will usually be seen to be one of five shapes:
  (a) Round–(when the widest part is central and the cell roundish in shape).
  (b) Ovate–(when the widest part of the cell is nearest the base of the apple).

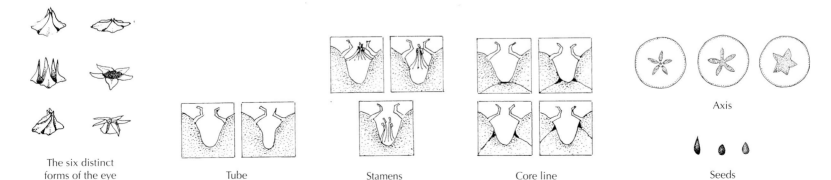

The six distinct forms of the eye   Tube   Stamens   Core line   Axis   Seeds

(c) Obovate–(when the widest part of the cell is nearest the eye of the apple).

(d) Elliptical–(when the widest part of the cell is central but the cell is elliptical instead of round).

(e) Lanceolate -(when the cell tapers to a point at both ends).

Tufted In some cultivars the inside of the seed cells have some white woolly ribboning, for example Cornish Gilliflower and Lane's Prince Albert.

Axis If a horizontal section is cut, the attachment of the cells to the axis can be seen. If the cells are symmetrical they are axile, whether (a) open or (b) closed. If the cells are asymmetrical they are (c) Abaxile.

Seeds They can be (a) Acuminate–(long and sharply pointed), (b) Obtuse–(short and blunt), (c) Acute–(somewhere between the previous two).

Flesh Variations in the colour and texture of the flesh are usually constant within a cultivar. The taste of the fruit is subjective and can again vary according to soil and climate.

Aroma This usually indicates the scent of the fruit before it is cut, unless otherwise stated.

TREE CHARACTERISTICS

Flowers

The grouping numbers refer to the flowering time and are listed from 1–7 on the pollination table.

Triploid

Plants grow by the multiplication of cells brought about by cell division. When the cell divides, thread-like structures can be seen in the nucleus which are called chromosomes. The number of chromosomes found during cell division is characteristic of a species. The majority of apple cultivars grown today contain thirty-four chromosomes and are known as diploid. When the cell divides, the chromosomes separate into two equal sets of seventeen, one set donated by the male parent, the other by the female. In the case of a triploid the number of chromosomes is fifty-one and cannot be evenly divided and it is found that the larger set is usually donated by the female and that the pollen is sparse. Triploids are therefore not used as pollinators, so it is necessary to plant three cultivars within the same group, one to pollinate the triploid and one to pollinate the pollinator and triploid.

Leaves

There are two types of leaf, the leaves attached to the spur and the leaves of the new growth shoots. It is the former that is described. The leaf shapes are divided into six categories:

(a) Oval
(b) Broadly oval
(c) Narrow oval
(d) Acute
(e) Broadly acute
(f) Narrow acute.

The teeth along the leaf margins may be

(a) Serrate–(toothed in a saw-like manner)
(b) Bi-serrate–(when the serrations are again serrated)
(c) Crenate–(with rounded convex teeth)
(d) Dentate–(with vertical rather than angled teeth).

The surface of the leaf may be flat or undulating and the sides can be upward-folding or downward-folding. The undersides of the leaves are variably covered with pale downy hairs. With some cultivars the leaves tend to hang in a downward drooping fashion and that is referred to as downward-hanging.

GROWTH

Different cultivars vary in their habit of growth and if reasonably left to assume their natural shape will fall into roughly six categories.

(a) Upright
(b) Uprght-spreading,
(c) Spreading,
(d) Wide-spreading
(e) Compact
(f) Weeping.

Cultivars also differ in their habit of bearing fruit and it is important to ascertain the habit before pruning.

Spur system Trees that produce their fruit buds on two year old wood and on spurs on the older wood. Spurs are short jointed lateral branches carrying fruit buds. Each flower bud contains four or five flowers that should produce fruit. Behind that bud a new bud will form and the growth will continue in that way until after some years a multiple spur system will have formed.

Tip-bearer A habit found in some cultivars of producing most of their fruit from the terminal fruit buds of shoots made the previous summer. These cultivars will produce new spurs on the older wood. A good example is Irish Peach.

Spur and tip-bearer (partial tip-bearer) These cultivars will produce some fruits on the tips of shoots made the previous summer, as in tip-bearers but they also produce fruit on spurs made on the older wood, as in a typical spur-bearer. A good example is Discovery.

Oval   Broadly oval   Narrow oval   Acute   Broadly acute   Narrow acute

Upright   Upright-spreading   Spreading   Wide spreading   Compact   Weeping

EMNETH EARLY

An early Codlin-type cooking apple raised by William Lynn of Emneth in Cambridgeshire (UK) from Lord Grosvenor x Keswick Codlin and introduced by Messrs Cross of Wisbech. First recorded in 1899 with First Class Certificate from RHS. It later became known commercially as Early Victoria. It makes a hardy, spur-bearing, compact tree of moderately weak vigour and upright habit. The cropping is heavy with a biennial tendency and fruits can be small unless thinned. Season late July to mid August, picking time late July and early August. The fruits have a superb flavour, sub-acid and sweet, yet sharp when cooked, does not need sugar and cooks to a fluff. Fruits have almost no aroma.

SIZE Medium, though rather variable, 64–61mm (2^1/2 x 2^3/8")
SHAPE Round-conical to oblong-conical. Can be lop-sided. Irregular with prominent ribs.
SKIN Yellowish-green becoming fairly bright greenish-yellow. Frequently a hair-line present. Lenticels small whitish-green dots which become larger towards the base and into cavity. Skin smooth and dry.
STALK Medium thick (3mm) and quite long (23mm). Extends well beyond base.
CAVITY Fairly wide and shallow.
EYE Small. Closed. Sepals quite large and long, erect and held tightly together with some tips reflexed. Downy.
BASIN Shallow and quite small. Well puckered with trace of beading.
TUBE Cone-shaped.
STAMENS Marginal.
CORE LINE Basal.
CORE Median. Axile.
CELLS Elliptical. Tufted.
SEEDS Abundant. Acute. Plump and oval.
LEAVES Medium to fairly large. Oval to broadly oval. Crenate. Medium thick. Flattish and slightly undulating. Light yellowish-green with undersides downy.
POLLINATION GROUP 3

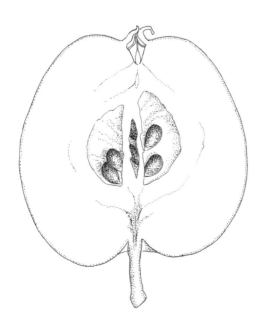

GLADSTONE

An attractive, very fragrant, very early dessert apple, season late July to mid August. It is deliciously sweet and juicy when picked and immediately devoured but should not be kept a moment longer as it becomes soft and dry. The flesh is white tinged green, soft and coarse-textured. It was introduced in 1868 by Mr. Jackson of Blakedown Nursery in Worcestershire (UK) as Jackson's Seedling, said to be a chance seedling originating about 1780. In 1883 it received a First Class Certificate from the RHS after which it was re-named Mr. Gladstone, after the Prime Minister. The 'Mr' is now generally omitted. It makes a moderaely vigorous, spreading tree which is a spur and tip bearer. It crops well but can be biennial. Picking time late July and early August.

SIZE Medium, 63 x 54 mm ($2^1/2$ x $2^1/8$")
SHAPE Round-conical. Rounded at base. Large well rounded ribs with one sometimes larger. Irregular. Can be lop-sided. Can be five-crowned at apex.
SKIN Variable in colour. Pale yellowish-green, partly to almost completely covered with deep red or brownish-red flush. Rather inconspicuous stripes of brownish-crimson. Some fruits can have a much more streaked appearance without the dense flush. Occasionally a bold wide stripe of yellow skin colour can show through the flush. Lenticels conspicuous as grey-brown russet dots on flush or purple-brown dots on green skin. Skin smooth becoming greasy after picking.
STALK Medium thick (3mm). Variable length (9–16mm). Protrudes beyond base.
CAVITY Fairly narrow and fairly shallow. Usually some golden-brown slightly scaly russet present which may come out on to the base.
EYE Medium size. Closed or slightly open. Sepals erect convergent.
BASIN Fairly deep and rather narrow. Ribbed. Russet free. Slightly downy.
TUBE Cone-shaped.
STAMENS Median.
CORE LINE Median towards basal.
CORE Median. Axile, open.
CELLS Ovate.
SEEDS Rather small. Acute. Light or golden brown. Very plump, rather rounded and pointed.
LEAVES Small to medium. Broadly oval. Serrate. Medium thick, Slightly upward-folding and undulating. Mid grey-green. Undersides very downy.
POLLINATION GROUP 4

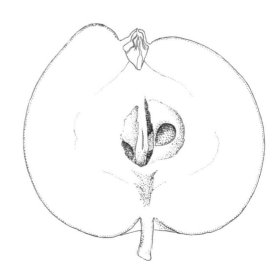

STARK'S EARLIEST

A late July to mid August dessert apple of American origin, discovered at Orofino, Idaho in about 1938 by Douglas Bonner. It was assigned to Stark Bros. Nurseries & Orchards Co. of Louisiana, Missouri and introduced in 1944. The parentage is unknown. The fruit ripens at about the same time as Beauty of Bath but remains in good condition for longer. It crops heavily though the fruit is apt to be small and the early flowering makes it unsuitable for colder areas. Picking time early August. It makes a moderately vigorous, upright-spreading tree that produces spurs freely and is resistant to scab. This variety has been superseded by Discovery commercially and is frequently listed under the name of Scarlet Pimpernel. The fruit is of quite good flavour, fairly juicy, crisp but tender with white fine-textured flesh. The skin is very tender.

SIZE Medium-small. 57 x 45mm (2¼ x 1¾")
SHAPE Variable. Flat-round to round-conical, sometimes conical. Irregular. Slightly lop-sided. Hint of well-rounded ribs, sometimes showing at base.
SKIN Very pale creamy-white. Flush can be a light dusting of pinkish-crimson or the fruit can be partly or almost completely flushed with bright crimson fading at the edges to milky-pink. Lenticels very distinctive on flush as large pale ochre dots with a deep crimson surround and on the ground colour as large russet dots with a greenish or pink surround. Skin smooth and dry.
STALK Fairly slender (2.5mm). Medium length (18mm). Protrudes beyond base.
CAVITY Medium width and depth. Streaked with greenish-ochre and fine golden-brown scaly russet which usually streaks out over base.
EYE Fairly large. Sepals broad, erect convergent, rather pushed together with tips well reflexed.
BASIN Rather narrow. Medium to shallow. Puckered and slightly downy.
TUBE Rather long funnel-shaped.
STAMENS Marginal.
CORE LINE Median towards marginal.
CORE Median. Axile open or abaxile.
CELLS Round or triangular.
SEEDS Plump. Obtuse. Slightly curved.
LEAVES Medium to smallish. Acute. Serrate. Medium thick. Flat. Mid to lightish green. Undersides moderately downy.
POLLINATION GROUP 1

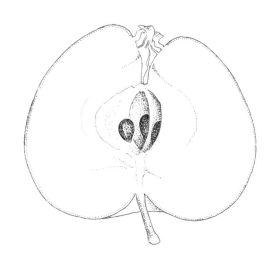

LAXTON'S EARLY CRIMSON

This very early, late July to mid August, dessert apple was raised in 1908 by Laxton Brothers of Bedford (UK) from Worcester Pearmain x Gladstone. It was introduced in 1931. It makes a spreading tree of weak to moderate vigour and is a partial tip-bearer. The cropping is moderate and the fruits should be eaten straight from the tree or within a few days of picking. Picking time late July and early August. The fruits are sweet, rather dry with a pleasant flavour and are very sweetly scented. The flesh is greenish-white, soft and fine-textured.

SIZE Medium, 67 x 63mm (2⅝ x 2½").

SHAPE Conical or long-conical. Rather waisted near apex. Large well-rounded ribs which end in a very pronounced ridge around the eye. Can be lop-sided but usually symmetrical. Fairly regular.

SKIN Dirty yellowish-green becoming dull pale greenish yellow. Three quarters to almost completely covered with deep crimson flush, paler on the shaded side and slightly striped and mottled over the yellow skin. There can be some short broken stripes of purplish-crimson on the paler flush otherwise no stripes. There may be some scarf skin at base. Lenticels are tiny crimson dots noticeable only on yellow skin. Occasionally a few small ochre russet surface patches.

STALK Fairly slender to medium (2.5–3mm) sometimes thickening towards base. Fairly long (17–21mm) protruding well beyond base.

CAVITY Shallow and fairly narrow. Yellowish green with some scaly golden-brown russet which can streak out over base.

EYE Medium size. Closed. Sepals connivent, fairly long and tapering, sometimes separated at base.

BASIN Narrow and deep with well pronounced crown which can look like large beads. Some scaly ochre-brown russet may be present.

TUBE Cone-shaped

STAMENS Median.

CORE LINE Sometimes two, median and marginal.

CORE Median to slightly sessile. Axile.

CELLS Round to roundish obovate, or roundish ovate.

SEEDS Small, fairly plump. Acute.

LEAVES Medium to small. Narrow acute. Serrate. Thick and leathery. Very upward-folding. Not undulating. Mid green. Undersides downy.

POLLINATION GROUP 2

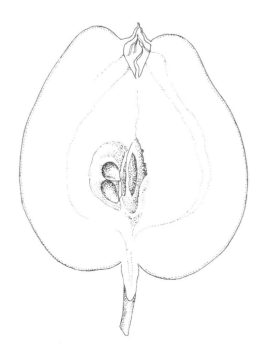

BEAUTY OF BATH

Beauty of Bath used to be one of the most important very early dessert apples grown commercially. It is picked in early August. The tree has a tendency to drop the fruit before it is fully ripe, therefore straw used to be placed beneath the 'Bath' trees in order to lessen the damage to the falling fruit. It was introduced in 1864 by Mr. George Cooling of Bath,(UK) who states that it is a Juneating seedling originating about the mid 1800s near Bath. The tree is slow to come into bearing and can be irregular in cropping due to its early flowering. It makes a hardy, moderately vigorous, spreading tree that produces spurs freely. It is fairly resistant to scab. The fruits have a full flavour, sweet with a slightly acid tang. The flesh is creamy-white, sometimes tinged pink near the flush, rather coarse-textured, juicy and soft. They are very sweetly aromatic with a fruity scent.

SIZE Medium. 63 x 51mm (2$^{1}/_{2}$ x 2").
SHAPE Flat-round. Regular. Sometimes a hint of ribs, usually noticeable at apex and occasionally at base. Symmetrical or lop-sided.
SKIN Pale whitish-green to yellow. Variably flushed and dotted with bright red. Traces of short broken stripes of brownish-crimson. Numerous large yellow lenticels give the flush a very mottled appearance. Usually some scarf skin at base which can extend over flush on cheeks. Some grey-brown russet streaks and patches. Skin smooth and greasy.
STALK Stout (4mm) and short (9–12mm). Usually within cavity or protruding slightly beyond.
CAVITY Fairly deep and wide. Scarf skin within cavity. Sometimes some fine grey-brown russet but often russet free.
EYE Can be almost closed with broad-based sepals erect and pressed together, even connivent, but often the tips are well reflexed leaving the eye half open and the stamens visible.
BASIN Wide and fairly deep. Usually slightly ribbed. Sometimes some fine grey-brown russet. Downy.
TUBE Funnel-shaped
STAMENS Median towards marginal.
CORE-LINE Median
CORE Median. Axile.
CELLS Round.
SEEDS Obtuse. Plumpish. Straight not curved.
LEAVES Medium size. Broadly oval. Crenate or bluntly serrate. Medium thick. Flat. Dark blue-green. Undersides not downy.
POLLINATION GROUP 2

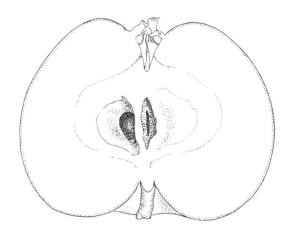

GEORGE CAVE

An early to mid August dessert apple raised in 1923 in Essex (UK) by Mr George Cave of Dovercourt, and said to be a chance seedling. It was acquired by Messrs Seabrook & Sons Ltd of Boreham, Essex but not named until 1945. The tree is fairly hardy, upright-spreading in habit, spur-bearing and of moderate vigour. The cropping is good though the fruit quickly drops. Picking time early to mid August. The fruits have no particular distinction: they are crisp, juicy and rather acid and the skin is a little tough. They are slightly sweetly aromatic. The flesh is creamy-white, sometimes tinged green under the skin, slightly soft, fine-textured and juicy.

SIZE Medium small, 57 x 51mm (2$^{1}/4$ x 2")

SHAPE Round-conical. Trace of well-rounded ribs. Regular. Symmetrical or slightly lop-sided.

SKIN Pale yellowish-green becoming yellow. Half or more covered with sparse red flush with some broken crimson-red stripes which can also appear on the green or yellow skin. There can be some scarf skin especially at the base which can radiate from the cavity as ragged dots and extend over flush, giving it a mottled appearance. Lenticels conspicuous as large pale ochre russet dots. There can be some small greenish-ochre russet patches. Skin smooth and dry.

STALK Medium thick (3mm). Quite long (15–20mm). Extends well beyond base.

CAVITY Rather shallow, medium width. Usually some scaly grey-brown russet which can extend over base. Often lipped.

EYE Medium size, partly open. Sepals broad based, long and slightly connivent, partly reflexed with some tips broken off. Downy.

BASIN Shallow. Medium width. Some puckering.

TUBE Cone-shaped.

STAMENS Median.

CORE LINE Median towards basal.

CORE Median. Axile.

CELLS Obovate.

SEED Acuminate. Fairly plump and fairly regular.

LEAVES Medium size. Acute to narrow acute. Serrate. Medium thick and leathery. Slightly upward-folding. Dark grey-green. Undersides very downy.

POLLINATION GROUP 3

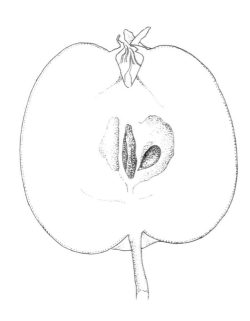

LADY SUDELEY

A very handsome second early dessert apple which was raised in England at Petworth in Sussex in 1849 by a cottager named Jacob and originally known as Jacob's Strawberry. The name is said to have been given by Lord Sudeley who, when having an orchard laid out with this variety, was reminded of a striking dress worn by his wife at Court. In 1884 the variety was given an RHS Award of Merit and it was introduced by George Bunyard & Co in 1885. It makes a compact tree of moderate vigour and is worth growing for the beauty of its fruit. It is a tip-bearer, the cropping is good and it is a useful early variety to grow in areas subject to late frosts because of its late flowering. Picking time mid August. The fruit has a good flavour with soft juicy creamy yellow flesh but quickly becomes dry and woolly with a slightly acid after-taste. The aroma is sweet and delicate, suggesting raspberries.

SIZE Medium 67 x 57mm (2⁵/8 x 2¹/4")

SHAPE Round-conical to oblong-conical. Flattened at base and often five-crowned at apex. Fairly distinct ribs. Can be lop-sided. Regular.

SKIN Pale greenish-yellow becoming golden yellow. Half to almost completely covered with bright orangey-red flush. Very prominent long broad stripes of bright crimson sometimes appearing over yellow skin. Occasionally some small russet patches. Lenticels very conspicuous as large greenish-brown or pale ochre russet dots. Can be some patchy scarf skin at base. Skin smooth and dry becoming greasy.

STALK Medium thick (3mm) and short (5–10mm.) Usually set within cavity.

CAVITY Medium width and fairly deep. Often green and always russetted with fine ochre or scaly brown russet which can streak over shoulder.

EYE Medium size closed or half open. Sepals erect convergent. Often stamens present.

BASIN Deep and fairly narrow. Prominently ribbed, often with a small amount of ochre-brown russet.

TUBE Cone or funnel-shaped.

STAMENS Almost marginal.

CORE LINE Median.

CORE Median. Axile.

CELLS Oval tending towards obovate.

SEEDS Numerous. Fairly large, plump. Acuminate.

LEAVES Medium size. Acute. Serate. Medium thick. Flat. Mid grey-green. Undersides very downy.

POLLINATION GROUP 4

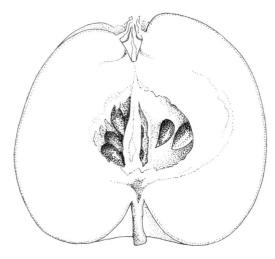

DISCOVERY

This second early mid August to mid September dessert apple has become very popular commercially. It was raised in 1949 by Mr Dummer of Langham in Essex (UK) from Worcester Pearmain and possibly Beauty of Bath. Grafts were bought by Jack Matthews of Matthews Fruit Trees Ltd in Thurston, Suffolk and named Thurston August. In October 1962 it was re-named Discovery and introduced in 1963. The trees are moderately vigorous upright-spreading and show a resistance to scab. They are fairly hardy and the blossom shows good tolerance to frost, making it suitable for growing in colder areas. It is a spur and tip bearer. It is rather slow to come into bearing and has a tendency to drop quantities of fruitlets from young trees but once mature crops heavily. Picking time mid August. The fruit has a remarkably long shelf life for an early dessert. The flavour is quite good and the flesh is creamy white sometimes tinged pink at flushed edges, fine-textured, crisp and juicy. It has almost no aroma.

SIZE Medium-small, 60 x 48mm (2³/8 x 1⁷/8″).
SHAPE Flat to flat-round. Well flattened at base and apex. Regular. Hardly any trace of ribs. Usually symmetrical.
SKIN Pale greenish-yellow becoming creamy-yellow. Half to three quarters covered with brilliant crimson-red flush. Some indistinct carmine stripes. Some scarf skin at base. Lenticels conspicuous yellow or pinkish russet dots on flush or less conspicuous tiny grey dots on yellow skin. Skin smooth and dry becoming slightly greasy.
STALK Stout (4mm) and short (10mm). Level with base or protuding beyond.
CAVITY Fairly wide. Medium depth. Lined with fine ochre-green and some light brown scaly russet, which can streak out a little over base.
EYE Medium size. Closed. Sepals erect convergent with tips reflexed or broken off. Downy.
BASIN Moderately shallow. Medium width. Slightly ribbed. There can be some small beads.
TUBE Cone-shaped.
STAMENS Median.
CORE LINE Basal.
CORE Median. Axile
CELLS Round.
SEEDS Acuminate. Fairly plump. Regular.
LEAVES Medium to small. Acute. Serrate. Fairly thick and leathery. Flat . Upward-folding. Mid green. Undersides downy.
POLLINATION GROUP 3

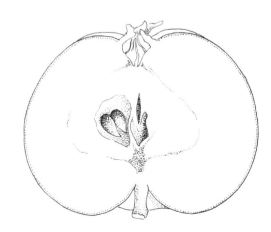

IRISH PEACH

This apple is supposed to be of Irish origin possibly from County Sligo, but exactly where it came from or who raised it remains a mystery. It was introduced into England in 1820 by Mr. John Darby of Addiscombe and Mr Robertson of Kilkenny. It is an attractive, fine quality, second early dessert apple, season late August to early September. If caught right it has a lovely rich vinous flavour with plenty of juice but it quickly goes soft and dry. It is best when eaten straight from the tree in late August and will not keep. The fruits have soft, fairly juicy, creamy-white flesh tinged very slightly green and are strongly scented. The tree is of moderate vigour, spreading in habit and is a pure tip-bearer forming few or no spurs. It is a rather gaunt looking tree. The cropping is irregular.

SIZE Medium 61 x 48mm ($2^3/8$ x $1^7/8$″)

SHAPE Round rather flattish to slightly round-conical. Fairly conspicuous broad ribs which can be rather angular. Some fruits rather flat-sided. Can be slightly five-crowned at apex. Irregular. Usually symmetrical.

SKIN Pale yellow tinged green with very mottled brownish-orange flush. Broken often broad stripes of milky-crimson. Lenticels very conspicuous on yellow skin appearing as grey-brown or greenish russet dots and small triangles, less conspicuous on flush where they appear more ochre-brown. Occasionally some patches of ochre-brown finely scaled russet. Skin smooth and rather greasy.

STALK Rather stout (3–4mm) and sometimes fleshy. Medium length (15mm).

CAVITY Medium to small and quite shallow. Usually some fine scaly brown russet radiating from stalk.

EYE Fairly small. Closed. Sepals connivent or erect convergent.

BASIN Wide. Medium depth. Ribbed and sometimes beaded.

TUBE Cone-shaped.

STAMENS Median.

CORE LINE Almost basal.

CORE Median. Axile slightly open.

CELLS Broad ovate or obovate. Tufted. Rather transparent looking.

SEEDS Roundish oval, acute. Straight not curved.

LEAVES Medium size, broadly oval. Bluntly serrate. Medium thick. Flat not undulating. Very slightly upward-folding. Mid blue-green. Undersides very downy.

POLLINATION GROUP 2

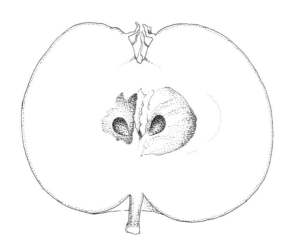

MILLER'S SEEDLING

Miller's Seedling is a second-early dessert apple raised in 1848 by James Miller, a nurseryman from Newbury, Berkshire (UK). Season late August and early September. It used to be widely grown for the markets and had a very high reputation in Berkshire. It received the Award of Merit from the RHS in 1906. It makes a moderately vigorous, upright-spreading tree and is a spur-bearer. Picking time mid August. The cropping is prolific with biennial tendency and the fruits can be small unless well thinned. They are easily bruised. They have a distinctive though not strong flavour being sweetly tangy and slightly scented. They are pleasant and refreshing with white fine-textured flesh which has a tinge of pink when fully ripe. The flesh is soft but crisp and very juicy. A red sport exists which is said to mature a week later.

SIZE Medium, 63 x 54mm (2^1/2 x 2^1/8 ″).
SHAPE Round-conical. Slight trace of ribs and sometimes rather flat-sided. Regular. Symmetrical or lop-sided.
SKIN Pale greenish-yellow becoming cream. Flush can be a blush of pale orangey-pink or can be bolder and brighter mottled milky-pink with short broken bright milky-red stripes. Lenticels inconspicuous whitish dots. Skin smooth and slightly greasy.
STALK Slender (2mm) and long (18–26mm). Protrudes well beyomd base.
CAVITY Deep and wide. Regular. Either russet free or with some fine ochre-brown russet which can be overlaid with grey-brown scaly russet streaking from the stalk.
EYE Small. Closed. Sepals quite large, broad based, connivent with tips reflexed. Very downy.
BASIN Medium width, shallow with the eye sometimes almost standing on top of the fruit. Ribbed and puckered, sometimes beaded.
TUBE Cone-shaped.
STAMENS Marginal.
CORE LINE Median.
CORE Median. Axile wide open or abaxile.
CELLS Elliptical to ovate.
SEEDS Flat and roundish. Straight and acute.
LEAVES Medium to small. Broadly oval. Serrate. Medium thick. Very slightly undulating. Mid yellowish-green. Undersides downy.
POLLINATION GROUP 3

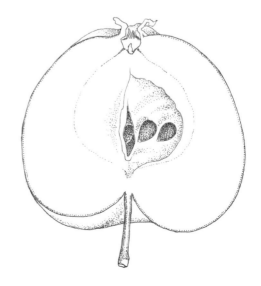

DUCHESS'S FAVOURITE

A very pretty second early dessert apple, season late August and September, raised in Addlestone in Surrey (UK) by a nurseryman named Mr Cree somewhere between the late 1700s and early 1800s. It was named in response to the acclaim with which it was received by the then Duchess of York. It used to be grown in the Kentish orchards for the London markets. The trees are moderately vigorous, upright-spreading spur-bearers and very fertile. Picking time late August. The fruits are sweet with a slight 'bite' possessing a pleasant though not strong flavour. The flesh is white tinged red between skin and core line, firm but tender and slightly coarse textured. The skin is rather tough and chewy. They have no aroma but have a slight acid scent when cut.

SIZE Medium-small. 58 x 51mm (2¼ x 2").

SHAPE Flat-round. Flattened at base and apex. Occasional trace of ribs. Regular. Symmetrical.

SKIN Pale greenish-yelow becoming pale yellow. Half to almost completely covered with bright red flush varying from light rather orangey-red on shaded side to crimson red on sunny side. Some faint short broken stripes more noticeable on paler flush. Large pale ochre lenticels becoming smaller, whiter and more numerous towards apex. Skin shiny and smooth becoming greasy.

STALK Slender (2mm) and fairly short (12mm). Level with base or protruding slightly beyond.

CAVITY Medium width and moderately deep. Lined with fine golden russet which streaks out over base. There can be some scarf skin appearing as pinky-white streaks or mottling over the flush at the base.

EYE Small to medium. Open. Sepals long if not broken off, erect with tips well reflexed. Some stamens usually visible.

BASIN Shallow. Wide. Puckered and slightly ribbed.

TUBE Mostly cone-shaped (Robert Hogg in *The Fruit Manual* found it funnel-shaped).

STAMENS Median towards marginal.

CORE LINE Median. Sometimes red.

CORE Median. Axile.

CELLS Round, slightly ovate or slightly obovate.

SEEDS Accuminate to acute. Numerous. Rather angular and slightly curved. Plump.

LEAVES Medium to small. Acute. Bluntly and broadly serrate. Medium thick. Upward folding and slightly undulating. Mid grey-green. Undersides not downy.

POLLINATION GROUP 3

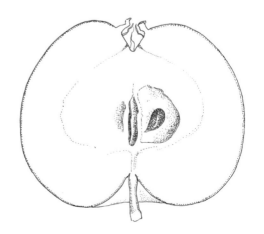

DEVONSHIRE QUARRENDEN

The origin of this very old second early dessert apple is uncertain. According to Bunyard in *A Handbook of Hardy Fruits* this apple is mentioned by Worlidge in his *Vinetum Britannicum* in 1678 and probably takes its name from Carentan, an apple growing district in France. It is also mentioned in *The Complete Planter and Cyderist* published in 1690. Some say it is a native of Devon where it was certainly a favourite and extensively cultivated. It thrives in the West Country as it will tolerate rain and wind. The season is late August to early September. The tree is upright-spreading, moderately vigorous and a spur-bearer. It can be rather susceptible to scab. The cropping is moderate to heavy. Picking time late August. The attractive fruits are sweet, crisp and juicy with a lovely fruity flavour. The flesh is greenish-white and slightly coarse.The core-line and flesh can sometimes be tinged red.

SIZE Small. 51 x 38mm (2 x 1¹/₂″).
SHAPE Flat-round. Rounded at base and apex. Some ribs present which can be rather angular. Irregular. Symmetrical or lop-sided.
SKIN Yellowish-green. Three quarters to completely flushed with crimson-brown to dark brownish-purple. Definite patches of green skin can remain where the skin has been shaded by another fruit or leaf. Lenticels inconspicuous pinky-grey dots on flush, conspicuous purple dots on green skin. Some rather indistinct brownish-crimson stripes noticeable at flush edges. Skin smooth and greasy.
STALK Medium thick (3mm) Medium length (15mm). Protrudes beyond base.
CAVITY Fairly wide. Medium to fairly shallow. Regular. Lined with some fine pale grey-brown russet.
EYE Large. Closed. Stamens long, connivent with tips well reflexed. Very downy.
BASIN Wide and shallow. Sometimes the eye is almost sitting on top of the fruit. Puckered and lumpy with irregular shaped beads.
TUBE Long funnel-shaped, extending up to core.
STAMENS Marginal.
CORE LINE Median.
CORE Median, Axile open, occasionally abaxile.
CELLS Ovate.
SEEDS Obtuse. Numerous. Regular.
LEAVES Medium size. Acute. Sharply or bluntly serrate. Medium thick. Upward folding not undulating. Light grey-green. Very downy.
POLLINATION GROUP 2 Can be Biennial.

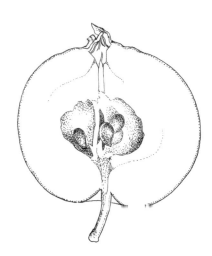

LORD GROSVENOR

The origin of this second early Codlin-type cooking apple is unknown. It is recorded as 'new' in *Scott's Orchadist,* a catalogue of fruits cultivated at Merriott in Somerset c.1873 and was once extensively grown in gardens and orchards. It was superseded by Early Victoria. The season is mid August to mid October. It makes an upright-spreading tree of weak vigour and is a spur-bearer. The cropping is very heavy. Picking time mid August. The trees are fairly hardy and suitable for colder areas. The fruits are acid with fairly good flavour and they break up completely when cooked though not to a fluff. The flesh is white tinged green, fine textured and firm and fairly juicy. No aroma.

SIZE Medium-large, 73 x 64mm (2^{7}/$_{8}$ x 2^{1}/$_{2}$").
SHAPE Round-conical to conical. Definitely ribbed. Irregular and angular in shape. Frequently lop-sided.
SKIN Pale yellowish-green becoming yellow. Occasionally a tinge of pinkish-brown otherwise no flush or stripe. A hair line may be present. Some scarf skin at base. Lenticels conspicuous whitish-green dots which become larger towards base and frequently enter stem cavity. Some pinkish-brown russet dots within the white dots. Skin smooth and dry.
STALK Medium thick (3mm) long (25–28mm). Protrudes beyond base.
CAVITY Wide and fairly deep. Scarf skin within cavity. Occasionally some grey-brown russet within. Usually slightly ribbed.
EYE Large. Closed. Sepals connivent and pressed well together with tips reflexed. Downy.
BASIN Small and shallow. Extremely pinched looking as though the skin has been gathered around the eye. Ribbed and beaded.
TUBE Long. Rather straight cone-shaped.
STAMENS Marginal.
CORE LINE Median.
CORE Median. Abaxile.
CELLS Elliptical or slightly ovate. (Hogg in *The Fruit Manual* found them ovate).
SEEDS Light golden-brown. Fairly plump. Acute.
LEAVES Medium size, broadly oval. Broadly and bluntly serrate. Fairly thick and leathery. Undulating. Mid green. Undersides slightly downy.
POLLINATION GROUP 3

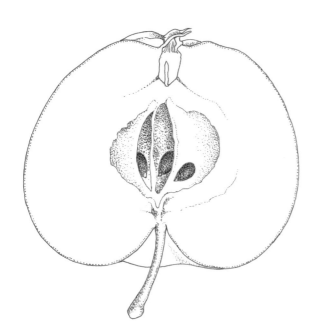

GRENADIER

Grenadier was brought to notice by George Bunyard of Maidstone in Kent (UK) and first recorded in 1862. It is thought to have been cultivated for many years prior to that although there is no record of its origin. Grenadier was introduced commercially in 1875 and received a First Class Certificate from the RHS in 1883. It makes a spreading tree of moderate vigour which spurs freely and crops heavily. Season August to October, picking time mid August. It is hardy and resistant to scab and canker but susceptible to capsid bug. It is suitable for growing in the north. The fruits are tangy and juicy with a superb slighly honey flavour. They cook to a fluff and are good for baking and purees. The flesh is white tinged green, fine-textured and juicy. Aroma almost nil.

SIZE Large, 83 x 63mm (3^{1}/4 x 2^{1}/2″).
SHAPE Round-conical rather flattish. Distinct well-rounded ribs. Sometmes flat-sided. Irregular. Slightly five-crowned at apex. Symmetrical or lop-sided.
SKIN Fresh pale green becoming pale yellowish-green. Some scarf skin mainly at base. Lenticels conspicuous as numerous white dots at apex and green or brown russet dots on cheeks. Skin smooth and dry.
STALK Stout (4mm) and short (10–12mm). Usually fleshy. Does not extend beyond base and is set deep within cavity.
CAVITY Wide and deep. Usually regular but can have a large rib extending into cavity making it uneven. A trace of grey-brown scaly russet can streak out over base.
EYE Small. Closed. Sepals connivent with tips reflexed. Fairly downy.
BASIN Narrow and fairly shallow. Much puckered and ribbed.
TUBE Cone-shaped.
STAMENS Median
CORE LINE Basal to median.
CORE Median. Abaxile.
CELLS Elliptical.
SEEDS Rather small. Mid brown. Straight. Acute.
LEAVES Medium size. Fairly broadly acute. Small bluntly pointed serrate. Fairly thick and leathery. Undulating. Mid green. Undersides downy.
POLLINATION GROUP 3

LAXTON'S EPICURE

A high quality second early dessert apple which was raised in 1909 by Laxton Brothers Ltd of Bedford (UK) from Wealthy x Cox's Orange Pippin. Season late August to mid September. It was awarded the Bunyard Cup in 1929 and 1932 and received the Award of Merit from the RHS in 1931. It makes a hardy, upright-spreading tree of moderate vigour and is a spur bearer. Picking time late August. The trees are hardy and frost resistant making them suitable for planting in colder areas. The fruits are sweet, juicy and refreshing and somewhat pear-flavoured though the skin is a little tough. The flesh is creamy-white sometimes tinged pink, rather coarse-textured, slightly soft but firm. The aroma is faint when uncut, vinous when cut. The tree tends to over-crop with small fruits.

SIZE Medium 63 x 54mm (2$^{1}/_{2}$ x 2$^{1}/_{8}$").
SHAPE Flat-round to round-conical. Regular in outline with very slight trace of ribbing. Symmetrical.
SKIN Pale yellowish-green. Quarter to three quarters flushed with dull brownish-red. Short broken stripes of dull brownish-crimson. Lenticels inconspicuous small pinky-grey or pale ochre-grey russet dots. Skin smooth and dry.
STALK Fairly slender (2.5mm), long (30–35mm). Extends well beyond base. Sometimes with a fleshy knob.
CAVITY Wide and deep. Regular. Usually russet free but occasionally some fine pale grey-brown russet at the base of the cavity.
EYE Medium size, closed or partly open. Sepals broad-based and connivent with tips reflexed.
BASIN Medium width and fairly shallow. Puckered and slightly ribbed.
TUBE Cone-shaped.
STAMENS Medan.
CORE LINE Median.
CORE Median. Axile.
CELLS Roundish or elliptical tending towards ovate.
SEEDS Quite large and numerous. Mid brown. Acute, fairly long but bluntly pointed. Angular.
LEAVES Medium size, broadly acute. Serrate to broadly serrate. Thick and leathery. Slightly upward-folding and undulating. Mid to dark grey-green. Undersides not downy.
POLLINATION GROUP 3

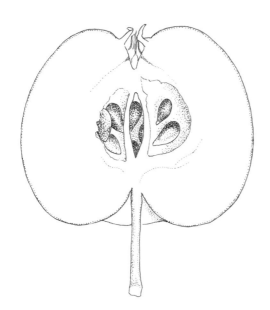

OWEN THOMAS

This second early dessert apple was raised in 1897 by Laxton Brothers of Bedford (UK). It was introduced in 1920. It makes a wide-spreading moderately vigorous tree which is a spur-bearer. The cropping is light to moderate. Picking time mid to late August. It has a very short season, late August and early September and for a brief moment the fruits are quite nice and aromatic if eaten straight from the tree but in two or three days they are over. The flesh is greenish-white, fine textured with some transparent green parts in the flesh, soft and fairly juicy. Aroma very slight, sweet yet slightly acid.

SIZE Medium-small, 57–51mm (2$^{1}/4$ x 2″).
SHAPE Flat-round. Rounded at base. Some large well pronounced ribs. Irregular. Can be slightly five-crowned at apex.
SKIN Yellowish-green becoming yellow. Quarter to three quarters flushed with orange-red which can either be sparse or in broad bands. Skin under or at edges of flush is golden. Some broken scarlet stripes. Lenticels indistinct small grey-green dots on green and yellow skin, pinkish-white dots on flush. There can be some fine grey-brown russet patches and some patchy scarf skin at base and on cheeks. Skin smooth and slightly greasy.
STALK Fairly stout (3.5mm) Short to medium (7 -15mm). Extends beyond base of fruit.
CAVITY Medium to rather shallow and narrow. Sometimes lipped. There can be some fine ochre-brown russet radiating from the stalk which can continue slightly over base.
EYE Fairly small. Tightly closed or slightly open. Sepals small, erect or convergent with tips reflexed. Sometimes stamens visible. Very downy.
BASIN Medium width and depth. Much ribbed and puckered. There can be some beading.
TUBE Funnel-shaped.
STAMENS Marginal.
CORE LINE Median.
CORE Median to slightly distant. Axile.
CELLS Elliptical tending towards ovate.
SEEDS Acute. Fairly plump. Bluntly pointed and straight.
LEAVES Small. Acute. Deeply cut serrate. Thick and leathery. Upward-folding and undulating. Mid blue-green. Undersides quite downy.
POLLINATION GROUP 2

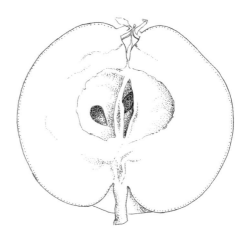

TYDEMAN'S EARLY WORCESTER

This McIntosh type second early dessert apple was raised at East Malling Research Station in Kent (UK) in 1929 by H.M.Tydeman from McIntosh x Worcester Pearmain. It was introduced in 1945 and named in 1963. The trees are moderately vigorous, very spreading with some long arching lateral branches. Season late August to mid September, picking time mid August. It bears most of the fruit on spurs and a small percentage on tips. The young tree will produce less fruit if the laterals are too severely pruned during the early years, otherwise the cropping is good. The fruits have a good rich flavour, sweet with a slightly acid bite. The flesh is white, fine-textured and juicy. The fruits are very sweetly scented.

SIZE Medium. 67 x 60mm. (2⅝ x 2⅜").

SHAPE Round to round-conical. There can be a trace of ribs and sometimes one prominent one, otherwise regular. Can be lop-sided.

SKIN Greenish-yellow becoming pale yellow. Half to almost entirely covered with crimson-red flush. Indistinct purplish-crimson stripes. On the shaded side it is often mottled and streaked with a pale grey-red with the yellow skin showing through. Lenticels numerous and reasonably conspicuous pale pinky-white dots on flush and pale yellowish-white dots on shaded side. Skin smooth and slightly greasy becoming more greasy if stored.

STALK Fairly slender (2.5mm) and fairly long (17–20mm). Extends beyond base.

CAVITY Fairly deep and rather narrow. Olive green lined with fine grey-brown finely scaled russet which streaks out a little over base.

EYE Small. Tightly closed. Sepals short, convergent with tips reflexed. Downy.

BASIN Medium depth. Rather narrow. Often beaded.Slightly puckered.

TUBE Cone-shaped.

STAMENS Median.

CORE LINE Median.

CORE Median. Axile open or abaxile.

CELLS Obovate.

SEEDS Acuminate. Straight. Fairly plump.

LEAVES Medium size. Acute. Serrate. Medium thick and leathery. Not undulating slightly upward-folding. Mid grey-green. Undersides downy.

POLLINATION GROUP 3

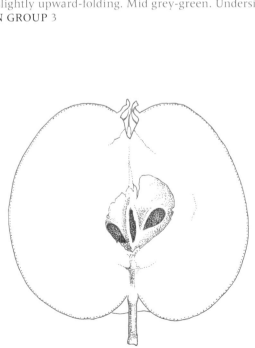

GEORGE NEAL

This high quality second early culinary apple was raised in 1904 by Mrs Reeves at Otford in Kent. It was introduced in 1923 by R. Neal and Sons of Wandsworth in London and in that year received an Award of Merit from the RHS. The high colour of this apple has given the impression that it is not of high quality which is a pity because the fruits have an excellent flavour, sweet yet a little acid. When cooked, the fruit becomes pale yellow and slices remain intact. The tree is of moderate vigour, spreading in habit and a spur-bearer. The cropping is good. Picking time late August and early September.

SIZE Large. 82 x 63mm (3¼ x 2½″).
SHAPE Flat-round. Broad at base and apex. Slight broad well-rounded ribs. Frequently lop-sided. Fairly regular.
SKIN Pale green becoming pale greenish-yellow. Flush orange-brown with short broken stripes and dots of bright red. There can be some patches of greenish-grey russet. Lenticels fairly conspicuous ochre-grey dots on flush and grey-brown russet or green dots on ground colour. Skin smooth and slightly greasy.
STALK Fairly slender (2.5mm) and short to medium length (10–20mm) Within cavity or protrudes slightly beyond.
CAVITY Narrow and deep. Lined with grey-brown scaly russet which can run concentrically round cavity and can come out over shoulder.
EYE Fairly large. Open. Sepals short and erect with some tips slightly reflexed. Slightly downy.
BASIN Wide and deep. Slightly ribbed. Can be some tiny specks of grey-brown russet.
TUBE Cone-shaped or slightly funnel-shaped.
STAMENS Median.
CORE LINE Median towards basal.
CORE Median, slightly distant. Abaxile.
CELLS Roundish.
SEEDS Acute. Fairly plump. Straight.
LEAVES Medium size. Acute. Bluntly serrate. Medium thick. Slightly upward-folding. Flat. Mid green. Undersides downy.
POLLINATION GROUP 2

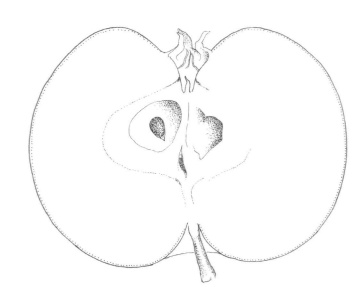

MERTON KNAVE

This very attractive and brightly coloured mid season dessert apple was raised in 1948 at the John Innes Institute (UK) by M.B.Crane. It was developed from Laxton's Early Crimson x Laxton's Epicure. It was named Merton Ace in 1968 but re-named Merton Knave in 1970. It makes a moderately vigorous, round-headed tree which is spreading and rather weeping in habit. It is a partial tip-bearer. The tree is hardy and the cropping is good although the fruit is not of particularly high quality. Picking time mid September. The flavour is pleasant but not strong, sweet with a slightly acid tang. The flesh is creamy-white, slightly coarse textured and can be tinged pink just beneath the skin. It is quite tender and fairly juicy. The aroma is strong and fruity.

SIZE Medium small, 58 x 48mm (2¼ x 1⅞").
SHAPE Round. Sometimes flattened at base and apex. Occasional indistinct well rounded broad ribs but frequently no ribs. Regular. Usually symmetrical.
SKIN Greenish-yellow. Three quarters to almost completely covered with brilliant red flush with some areas of crimson red. Indistinct broken stripes of carmine which are more noticeable on shaded side. Lenticels conspicuous small crimson or brown dots on shaded side, blackish purple dots on flush. There can be some small ochre russet patches. Skin smooth becoming shiny and greasy if stored.
STALK Oval rather flattened (3.5 x 2.5mm). Length medium to very long (12–30mm). Extends beyond base.
CAVITY Wide. Medium depth. Some fine greenish-ochre russet radiating out in thin streaks.
EYE Medium size slightly open. Sepals connivent with tps well reflexed. Some stamens visible. Downy.
BASIN Medium width and depth. Sometimes beaded. Slightly ribbed and puckered.
TUBE Cone-shaped.
STAMENS Median to marginal.
CORE LINE Basal.
CORE Median. Axile.
CELLS Round.
SEEDS Numerous. Angular. Acute. Fairly plump.
LEAVES Medium to small. Narrowly acute. Bluntly serrate. Medium thick. Very upward-folding, slightly undulating. Dark grey-green. Downy.
POLLINATION GROUP 3

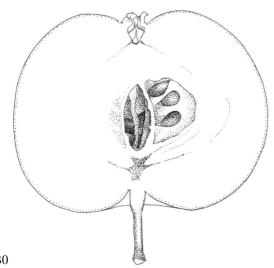

SCRUMPTIOUS

This is a new dessert variety developed by Hugh Ermen from Starkspur Golden Delicious and Discovery. It was introduced in 1980. The colour and the white flesh is typical of Discovery but the fruits of Scrumptious are larger. The trees are self-fertile, disease resistant and suitable for colder as well as wetter areas. The trees are moderately vigorous, upright spreading and spur bearing and the cropping is heavy even on young trees. The season and picking time is September. The fruits colour up early and are often attacked by birds The flesh is white, fine-textured, firm but soft and juicy. The flavour is rich and complex with a hint of aniseed. There is no aroma.

SIZE Medium-large. 70 x 60mm (2⁷/₈ x 2³/₈").
SHAPE Round-conical. Large well rounded ribs, can be a bit flat-sided. Symmetrical or slightly lop-sided. Rather irregular. Very slight trace of crowns.
SKIN Pale yellow. Half to completely covered with brilliant red to crimson flush. Can be distinct shapes of yellow in flush where fruit shaded by leaf. Broken stripes and patches of deeper crimson, less distinct on deeper flush. Tiny pale ochre or green lenticels. There can be some small pale red patches on the deep red flush like pale broken stripes. Skin smooth and dry.
STALK Slender to medium (2–3mm).Fairly long (22mm).Set at a slight angle. Extends well beyond base.
CAVITY Medium depth. Fairly narrow. Greenish ochre russet within.
EYE Fairly small. Closed or slightly open. Sepals erect convergent with tips slightly reflexed or broken off.
BASIN Quite shallow. Often beaded. Russet free. Downy.
TUBE Long funnel-shaped. Can extend into core.
STAMENS Median.
CORE LINE Median.
CORE Median.
CELLS Obovate. Axile open.
SEEDS Acute. Sharp-sided. Dark brown.
LEAVES. Acute. Serrate. Dark olive-green. Medium thick. Mostly flat. Slightly upward-folding. Undersides slightly downy.
POLLINATION GROUP 3. Self fertile.

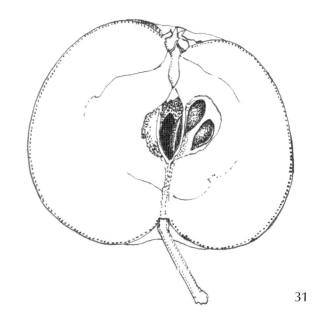

KATY

Correctly known as Katja, this second early dessert apple was raised in Sweden in 1947 by the Fruit Breeding Institute at Balsgard from James Grieve x Worcester Pearmain. The season is September to early October. It was selected in 1955, introduced in 1966 and named in 1968. It is quite widely grown commercially as it is fairly hardy, a good pollinator and more vigorous than Worcester Pearmain. It makes an upright-spreading tree which spurs freely. The setting is heavy and requires thinning in most years. Picking time early September. This apple does not taste quite as good as it looks, the flavour is fair, slightly acid and refreshing but the skin is rather tough. The flesh is white tinged green, fine textured and juicy with no aroma.

SIZE Medium, 66 x 60mm (2⁵/8 x 2³/8").

SHAPE Conical. Base rounded, tapering evenly towards apex which can be slightly five-crowned. Can be lop-sided. Some fruits faintly ribbed. Regular.

SKIN Very pale greenish-yellow becoming pale primrose yellow. Three quarters to almost completely covered with bright crimson-red flush. Rather indistinct short broken stripes of crimson. Areas of silverish-white dusting appear on the flush, mainly at base and apex. Lenticels indistinct as tiny whitish or pale yellow dots. Occasionally some small ochre russet dots or patches. Skin smooth and dry and polishes to a fine shine. Becomes greasy if stored.

STALK Medium to stout (3–4mm). Long (21mm). Nearly always set at an angle.

CAVITY Medium depth and width. Even. Lined with fine greenish-ochre russet overlaid with some fine scaly grey-brown russet which can streak a little over base.

EYE Fairly small. Closed. Sepals narrow, erect convergent with some tips slightly reflexed. Stamens can be visible.

BASIN Rather narrow and fairly shallow. Often beaded with large beads. Pinched looking. Usually russet free.

TUBE Cone-shaped.

STAMENS Median.

CORE LINE Basal, clasping.

CORE Median to sessile. Axile.

CELLS Roundish ovate.

SEEDS Acute, plump and slightly curved.

LEAVES Medium size. Narrow acute. Bluntly broad serrate. Medium thick. Upward-folding, not undulating. Mid green. Moderately downy.

POLLINATION GROUP 3

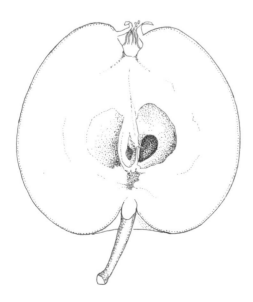

MERTON BEAUTY

A second early to mid season dessert apple from the John Innes Institute (UK). It was raised in 1932 by M.B.Crane from Elllison's Orange x Cox's Orange Pippin and released in 1962. It makes an upright-spreading tree of moderate vigour which produces spurs freely and, being late flowering, makes a useful garden cultivar. Picking time early September. The fruits are sweet and sharp with a distinct trace of aniseed which it gets from Ellison's Orange. It is an extremely good apple and the cropping is good. The flesh is creamy-white, fine-textured, crisp, firm and juicy. It has a very slight aroma.

SIZE Medium-small. 56–45mm (2³/₁₆ x 1³/₄").
SHAPE Flat-round to slightly conical. Usually no ribs but can have a slight trace. Regular and usually symmetrical.
SKIN Pale green becoming greenish-gold. Quarter to three quarters flushed with mottled brownish-red. Rather indistinct long broad stripes either of the same colour as flush or with a hint of crimson. Small patches and larger rather fuzzy areas of light brown or ochre russet chiefly towards base but also on cheeks. Skin smooth and dry.
STALK Medium thick (3mm) and long (20–27mm).
CAVITY Medium width and depth. Lined with light brown scaly russet which orbits the cavity rather than radiates from stalk. It can straggle out over shoulder.
EYE Medium size. Slightly open. Sepals very long or broken off. Convergent at base with three quarters of length reflexed. Fairly downy.
BASIN Fairly shallow, medium width. Slightly puckered.
TUBE Cone-shaped, or funnel shaped with a short funnel neck.
STAMENS Basal to medium.
CORE LINE Basal.
CORE Median. Axile
CELLS Obovate
SEEDS Rather large for size of fruit. Obtuse. Fairly broad, blunt ended. Not plump, rather angular and slightly curved.
LEAVES Medium size. Acute. Serrate. Thick and leathery. Slightly upward-folding, sometimes undulating. Mid green. Undersides not downy.
POLLINATION GROUP 5

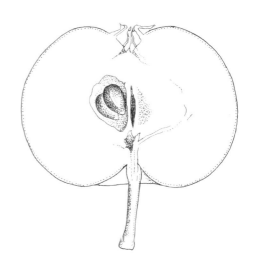

33

DELCORF

This is a most attractive and delicious dessert apple, the first good apple of summer which is ready to eat in August but will not keep much beyond September. It was raised in France by G Delbard of Alier (Auvergne) in 1956 as a cross between Golden Delicious and Stark Jonagrimes. It is also known as Delbarestivale and Delcorf Estivale. It was awarded the RHS Award of Garden Merit in 1998. The trees are fairly slow growing, fully hardy and late flowering. They produce heavy regular crops. The season is September to October, picking time early September. The flesh is creamy-white, fairly coarse-textured, firm but fairly soft and juicy. The fruit has a mixture of flavours, a honeyed sweetness with a hint of aniseed and a nice sharp bite making it refreshing. Aroma nil.

SIZE Large 80 x 80mm (3^1/8 x 3^1/8").
SHAPE Conical to oblong-conical. Hint of well rounded ribs with one rather more pronounced. Sightly five-crowned at apex, again one more pronounced. Irregular. Can be lop-sided.
SKIN Pale primrose-yellow becoming pale golden-yellow. Half to almost completely covered with broad broken bright red to crimson stripes and spots. Broad stripes of washed out red over yellow skin on shaded side. Stripes become thinner, darker and less broken on base and enter cavity. Among the stripes are some quite distinct yellow spots containing greenish-ochre lenticel, otherwise lenticels indistinct tiny russet dots. Skin smooth and dry.
STALK Long (22–26mm. Slender (2–3mm) Protrudes well beyond base.
CAVITY Deep and fairly narrow. Green with some pale grey-brown russet.
EYE Medium large. Open. Sepals long, erect with tips reflexed.
BASIN Narrow and fairly deep. Russet free. Trace of ribs. can be rectangular or round.
TUBE Cone-shaped.
STAMENS Median.
CORE LINE Median or towards basal.
CORE Median
CELLS Obovate. Axile.
SEEDS Acute. Mid brown. Flattish.
LEAVES Acute. Dark olive-green. Medium thick. Mostly flat. Slightly upward-folding. Serrate. Undersides slightly downy.
POLLINATION GROUP 3

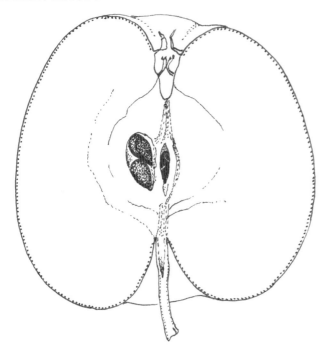

TIDICOMBE SEEDLING

This well flavoured mid to late dessert apple originated as a seedling at Tidicombe Farm, (now Tidicombe House), near Barnstable, Devon (UK). It was found in 1978 growing in the garden and thought to be a seedling from an apple core thrown from the house at some time. The fruit was brought to the attention of Kevin Croucher of Thornhayes Nursery by Ed Thoenburgh, the present owner of Tidicombe House and was introduced in 1997. The trees are modest of stature and compact in shape, resistant to scab and canker, and very wind tolerant. They are ideal for wetter and colder areas and for organic and pesticide free culture. They are spur bearers. The season is late September to late December/January, picking time September to early October. The fruits have a good rich flavour, sweet with a good acid balance and very slight aroma. The flesh is fine-textured, creamish- white, firm crisp and fairly juicy.

SIZE Medium 75 x 60–65mm (2⁷/8 x 2¹/2–2³/8").
SHAPE Round. Rounded at base. Very slight trace of well rounded ribs. Regular. Can be lop-sided.
SKIN Dull green becoming golden yellow. Slightly to three quarters covered with dots and flecks of scarlet with scattered short broken stripes of darker red. Quarter to half overlaid with fine pale grey-brown russet. This can be large areas or it can break and scatter over the underlying skin. Much golden or yellow skin with tiny dots and flecks of red showing through the russet. Skin dry and very slightly textured. Some large pale grey lenticels. An attractive golden looking fruit.
STALK Medium to long 20–30mm. Medium width 3–4mm. Protrudes slightly beyond base.
CAVITY Deep and fairly wide. Lined with pale grey-brown russet.
EYE Medium large. Wide open. Stamens visible. Sepals separated at base and fairly short, slightly reflexed or broken off.
BASIN Fairly narrow—as wide as the eye. Regular. Trace of ribs. Some fine grey-brown russet.
TUBE Funnel -shaped. Can be flattened out making it look almost cone-shaped.
STAMENS Median.
CORE LINE Median.
CORE Sessile.
CELLS Obovate or round. Axile slightly open.
SEEDS Acute. Dark brown.
LEAVES Very broadly acute. Small. Quite thin. Mid green. Serrate or bi-serrate. Leaves look quite distinctive on tree, small and broadly oval. Slightly undulating and upward-folding. Undersides slightly downy.
POLLINATION GROUP 2/3

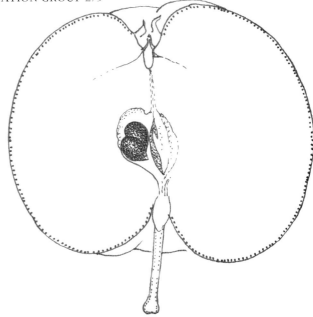

WORCESTER PEARMAIN

This is a well known second early to mid season dessert apple. It was raised by Mr. Hale of Swan Pool near Worcester (UK) and thought to be a seedling from Devonshire Quarrenden. It was introduced in 1874 by Messrs Smith of Worcester and in 1875 received a First Class Certificate from the RHS. It makes an upright-spreading, round-headed tree of moderate vigour and is inclined to tip-bearing, producing few spurs. It is hardy and reliable and is a heavy and regular cropper. Picking time early to mid September. It shows a resistance to mildew but is prone to scab. It is fairly frost resistant so suitable for colder areas. The fruits have a sweet and pleasant flavour which improves if the apples are left to ripen on the tree. Commercially they are picked too early resulting in inferior quality. The flesh is white, rather coarse-textured, firm and crisp yet tender and fairly juicy. It is very sweetly aromatic.

SIZE Medium. 64 x 61mm (2^1/$_2$ x 2^3/$_8$”).
SHAPE Conical to round-conical, occasionally long-conical. Symmetrical or lop-sided. Slight hint of ribs. Regular.
SKIN Pale greenish-yellow becoming pale yellow. Almost completely covered with brilliant red flush with indistinct long stripes of crimson-red. On shaded side, flush is pale red or pinkish red in long broad bands over the yellow skin with some brighter red stripes. Lenticels distinct as ochre russet dots. Skin smooth and greasy.
STALK Fairly slender (2.5mm) to fairly stout (3.5mm). Short to medium length (13mm).
CAVITY Rather narrow and deep with stalk deeply inserted. Lined with fine greenish-ochre russet which can streak or scatter over base.
EYE Fairly small. Closed. Sepals erect, connivent with tips reflexed. Fairly downy.
BASIN Shallow and narrow. Ribbed. Often beaded.
TUBE Long funnel-shaped.
STAMENS Median to marginal.
CORE LINE Median.
CORE Median. Axile.
CELLS Obovate sometimes slightly ovate.
SEEDS Quite large. Acuminate. Fairly plump.
LEAVES Medium size. Acute. Broadly serrate. Medium thick. Slightly upward-folding. Mid green. Undersides downy.
POLLINATION GROUP 3

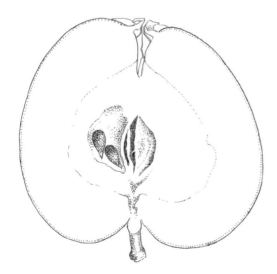

REVEREND W. WILKS

A large mid-season cooking apple which was raised from Peasgood Nonsuch x Ribston Pippin by Messrs Veitch of Chelsea, London. It was first recorded in 1904 by an Award of Merit from the RHS, was introduced in 1908, and received a First Class Certificate in 1910. The Rev. W. Wilks was the Vicar of Shirley in Surrey and Secretary of the Royal Horticultural Society from 1888–1919. He was devoted to horticulture and raised the first Shirley Poppy, *Papaver rhoeas*. Reverend W. Wilks makes a dwarf compact tree, suitable for small gardens. It has good disease resistance and is suitable for growing in the west. The trees are spur-bearers, produce spurs freely and the cropping is heavy with large fruits, though with a biennial tendency. Picking time early September. The fruits are sub-acid with a delicate aromatic flavour and cook to a pale yellow froth. The flesh is white, crisp, fine-textured and fairly juicy.

SIZE Very large, 89–76mm (3$^{1}/_{2}$ x 3").
SHAPE Round-conical to conical. Broad flattened base tapering to a flattened apex. Symmetrical. Slightly ribbed and fairly regular.
SKIN Pale whitish-green becoming primrose yellow. Quarter to half flushed with pale ochre or mottled pinky-red. There can be some broken scarf skin at base. A few definite broad broken stripes of bright milky-red. Lenticels inconspicuous sparse grey-brown russet dots. Skin smooth becoming greasy if stored.
STALK Fairly slender (2.5mm). Short to medium (10–16mm). Within cavity of slightly beyond.
CAVITY Wide and fairly deep. Slightly russetted with grey-brown slightly scaly russet.
EYE Medium size. Closed or partly open. Sepals erect or convergent. Stamens sometimes visible. Fairly downy.
BASIN Deep and medium wide. Distinctly ribbed.
TUBE Slightly funnel-shaped.
STAMENS Median.
CORE LINE Median.
CORE Median. Abaxile.
CELLS Roundish to elliptical.
SEEDS Acute. Fairly plump. Small.
LEAVES Large to medium. Broadly oval. Serrate. Medium thick. Slightly upward-folding and undulating. Dark green. Undersides slightly downy
POLLINATION GROUP 2

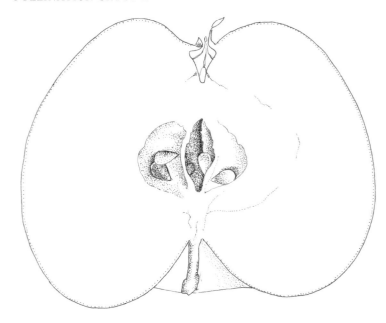

MERTON WORCESTER

A second early dessert apple from the John Innes Institute raised by M.B.Crane in 1914 from Cox's Orange Pippin x Worcester Pearmain. It was named in 1947 and received the Award of Merit from the RHS in 1950. It is rather susceptible to bitter pit. The trees are reasonably hardy, moderately vigorous and upright spreading in habit. They produce spurs very freely and the cropping is good and regular. Picking time early September. The season is September and October. The fruits have a good aromatic flavour, tasting slightly of pears, and are sweet and juicy. The flesh is creamy-white, firm, crisp and fine-textured.

SIZE Medium-small. 58–54mm (2¼ x 2⅛").
SHAPE Round-conical. Can be a slight trace of ribs and sometimes a little flat-sided. Fairly regular. Usually symmetrical. Can be lop-sided.
SKIN Yellowish-green becoming greenish-yellow. Half to three quarters flushed brownish-red, ochre-brown on less coloured fruits. Indistinct stripes of crimson-red. Lenticels inconspicuous as minute russet dots. Some small patches of pale grey-ochre russet. Skin smooth and dry.
STALK Medium (3mm). Length medium to long (12–30mm). Extends well beyond base of fruit.
CAVITY Medium width, medium depth with stalk deeply set. Usually some light brown russet running round cavity.
EYE Fairly small, closed. Sepals finely tapered, convergent with tips reflexed or broken off. Fairly downy.
BASIN Shallow. Medium width. Can be slightly ribbed. Usually five distinct beads.
TUBE Small, funnel-shaped.
STAMENS Median or towards marginal.
CORE LINE Basal.
CORE Median, Axile, slightly open.
CELLS Round.
SEEDS Acuminate. Long rather narrow and pointed. Very dark brown. Starved-loking not plump. Fairly straight.
LEAVES Medium to small. Acute. Bluntly pointed and broadly serrate. Thick and leathery. Upward folding, not undulating. Mid green. Undersides slightly downy.
POLLINATION GROUP 3

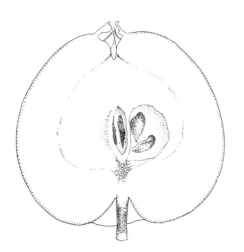

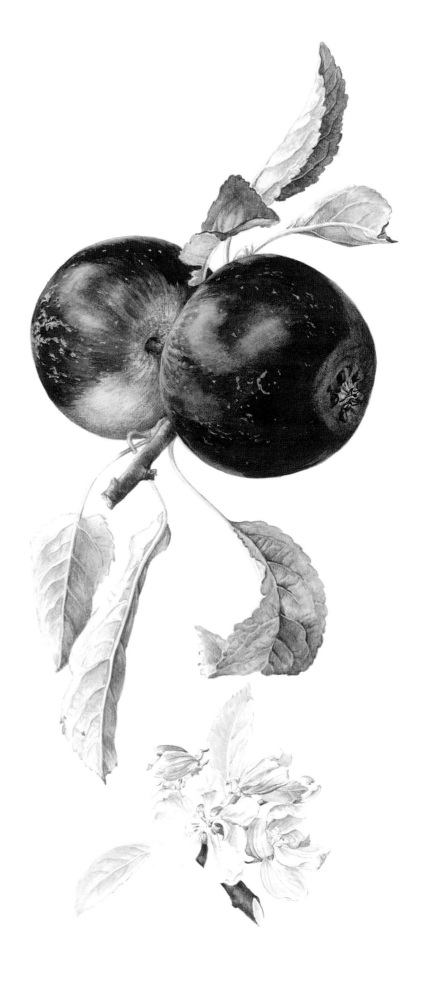

JAMES GRIEVE

A very popular second early dessert apple raised in Scotland by Mr. James Grieve of Edinburgh, open-pollinated from Pott's Seedling or Cox's Orange Pippin. Season September and October. It was introduced by Mr Grieve's employers, Dickson's Nurserymen, and first recorded in 1893. It received the Award of Merit from the RHS in 1897 and a First Class Certificate in 1906. Several coloured sports exist. The fruits bruise easily and may drop prematurely in warm districts. It prefers the North, disliking the humid West where it is prone to canker: otherwise it is hardy and adaptable. It makes a spreading, round-headed tree of moderate vigour. Picking time early September. The flavour is excellent, sweet with a nice acid balance but the skin is a little chewy. The fruits are acceptable as cookers in July and will keep until Christmas as a rather soft apple but with its flavour retained. The flesh is creamy-white, fine textured, soft and juicy. The fruits are sweetly scented.

SIZE Medium. 70 x 60mm (2³/₄ x 2³/₈").
SHAPE Round-conical to conical. Symmetrical or slightly lop-sided. Slight trace of ribs. Irregular. Slightly rounded or flattened at base.
SKIN Bright yellow-green becoming yellow. Variably speckled and striped with bright red over orange skin. Some small greenish-ochre russet patches. Lenticels conspicuous grey-brown russet dots. Skin smooth and greasy.
STALK Fairly slender (2.5mm) to medium (3mm). Long (26–33mm). Protrudes beyond base.
CAVITY Deep and wide. Greenish with some scaly grey-brown russet streaking out.
EYE Medium size. Closed. Sepals long, convergent and erect with tips reflexed. Sepals distinctly green with tips turning brown. Fairly downy.
BASIN Medium width to rather narrow. Medium depth. Slightly puckered with trace of ribs. Some russet lying concentrically round basin.
TUBE Funnel-shaped.
STAMENS Median.
CORE LINE Median towards basal.
CORE Median. Axile.
CELLS Obovate. Sometimes roundish.
SEEDS Quite large. Acute or acuminate. Fairly plump.
LEAVES Medium size. Acute. Crenate or bluntly serrate. Medium thick. Slightly upward-folding. Dark green. Undersides downy.
POLLINATION GROUP 3

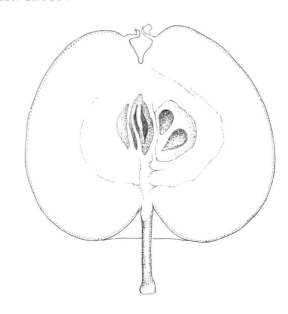

MERTON CHARM

A rather small but high quality mid season dessert apple. It was raised in 1933 at the John Innes Institute (UK) by M.B.Crane from McIntosh Red x Cox's Orange Pippin. It is an excellent garden apple for September and October. It makes a spreading, slightly weeping tree of moderate vigour which produces spurs freely. It is a good regular cropper and is fairly hardy. Picking time mid September. The fruits have a good flavour and are crisp, juicy and sweet. The flesh is creamy-white tinged slightly green near the skin and fine-textured. Almost no aroma.

SIZE Medium-small 57 x 47mm (2^{1}/4 x 1^{7}/8″).

SHAPE Round rather flattened to round-conical. A slight trace of ribs. Flattened at base and apex. Sometimes lop-sided.

SKIN Yellowish-green becoming golden yellow. Quarter to three quarters flushed brownish-red fading to ochre-brown. Overlaid with mottled and stripey brownish-crimson which can fuse together to give quite a bright crimson-red flush. Some scarf skin at base. Lenticels conspicuous on flush as tiny pinkish-white or pale yellow russet dots, pinkish-brown on green skin. There can be some small ochre russet patches. Skin smooth and dry becoming slightly greasy.

STALK Fairly slender (2.5mm) short to medium length (10–15mm). Can be fleshy. Within cavity or protruding slightly beyond.

CAVITY Quite narrow and fairly deep with stalk set well into it. Some fine ochre-brown russet may radiate from stalk which often occurs in patches on opposite sides of the stalk.

EYE Small. Closed or partly open with some stamens visible. Sepals small and narrow, erect or convergent with tips reflexed. Downy.

BASIN Narrow. Medium depth. Slightly ribbed.

TUBE Tiny, cone-shaped.

STAMENS Marginal.

CORE LINE Basal. Clasping. Rather indistinct.

CORE Median to slightly distant. Axile.

CELLS Roundish obovate.

SEEDS Acuminate. Fairly large for size of fruit. Mid brown. Not plump. Sightly curved.

LEAVES Medium size. Narrow acute. Quite broadly serrate. Thin. Upward folding not undulating. Light yellowish-green. Undersides downy.

POLLINATION GROUP 2

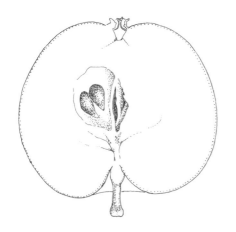

LAXTON'S FORTUNE

A mid season dessert apple raised in 1904 by Laxtons of Bedford (UK) from Cox's Orange Pippin x Wealthy. It was introduced in 1931, received an Award of Merit in 1932 and a First Class Certificate from the RHS in 1948. It is a fairly small compact tree of moderate vigour, season September and October, quite suitable for the small garden. It is generally listed as Fortune. The trees are fairly hardy. They show some resistance to scab but are prone to canker, otherwise they are fairly trouble-free. The cropping varies with area; in some it crops regularly and colours well but in others it tends to be biennial with poorly colour fruits. It is suitable for growing in the North and West. Picking time early September. The fruits are sweet with a good acid balance and aromatic flavour and a slightly sweetly scented aroma. The flesh is creamy-white, tender but firm and rather coarse-textured.

SIZE Medium. 67 x 61mm (2⅝ x 2⅜").

SIZE Medium. $67 \times 61mm$ ($2^5/8 \times 2^3/8$").
SHAPE Round, slightly conical. Flattened at apex, rounded at base. Symmetrical or lop-sided. Some indistinct ribs, more pronounced at apex. Five-crowned. Irregular.
SKIN Light green becoming yellow. Quarter to almost completely covered with broken stripes, speckles and mottled areas of bright red with the yellow or ochre-yellow skin showing through. Some patches of ochre-brown russet. Lenticels inconspicuous tiny grey russet dots. Skin smooth and dry.
STALK Fairly slender (2.5mm). Fairly long (25mm). Protrudes well beyond cavity.
CAVITY Deep. Medium width. Sometimes lipped. Lined with greenish-ochre and some slightly scaly brown russet which streaks and scatters over base.
EYE Medium size. Closed. Sepals broad and long, connivent and tightly pinched together. Sepals are of a distinctive green and tend to remain green rather than turn brown. Not downy.
BASIN Narrow and deep. Ribbed. Sometimes beaded. Some cinnamon-brown russet may radiate out over apex.
TUBE Cone-shaped, rather flattened at narrow end.
STAMENS Median.
CORE LINE Basal, clasping.
CORE Median to slightly distant. Axile.
CELLS Obovate.
SEEDS Obtuse. Round. Fairly plump. Straight.
LEAVES Medium size. Oval to broadly oval. Small bluntly pointed serrate. Medium thick. Slightly undulating. Mid green. Not downy.
POLLINATION GROUP 3

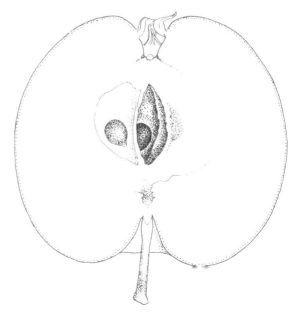

ELLISON'S ORANGE

This second early dessert apple was raised towards the end of the nineteenth centuary by the Rev.C.C.Ellison of Bracebridge in Lincolnshire (UK) and Mr. Wipf, the gardener at Hartsholme Hall, the home of Mr.Ellison's brother-in-law. It was raised from Cox's Orange Pippin x Calville Blanc. On Mr. Ellison's retirement he sold grafts of the tree for the financial benefit of Mr. Wipf and these were bought by Messrs. Pennell & Sons. This variety was first recorded in 1904. It was introduced and received an Award of Merit from the RHS in 1911 and a First Class Certificate in 1917. There are a number of red sports. The season is September and October, picking time mid September. The cropping is good though it tends to crop biennially. The trees are hardy and are suitable for growing in the North, but they do not like areas of high rainfall. The trees are prone to canker but are fairly resistant to scab. The fruits have a very rich and distinctive aniseed flavour and a fairly strong rather vinous aroma. The flesh is creamy-white tinged green, fairly fine-textured, crisp and juicy.

SIZE Medium 67 x 57mm (2⅝ x 2¼").
SHAPE Round, slightly conical, sometimes slightly oblong. Flattened at base and apex. Usually symmetrical. Slight trace of ribs. Regular.
SKIN Dull yellowish-green becoming yellow. Slightly to three quarters flushed with brownish-red and broad broken stripes of brownish-crimson. Some small russet scuffs and patches. Lenticels inconspicuous tiny green-grey russet dots. Skin smooth and slightly greasy, becoming more greasy when stored.
STALK Variable. Slender (2mm) and long (35mm) or medium -thick (3mm) and fairly short (15mm).
CAVITY Medium to shallow. Medium width. Lined with ochre and some fine light brown russet.
EYE Medium size. Tightly closed. Sepals erect, convergent or slightly connivent with tips reflexed or broken off. Fairly downy.
BASIN Medium width. Deep. Slightly ribbed.
TUBE Small, cone or slightly funnel-shaped.
STAMENS Median towards marginal.
CORE LINE Basal clasping.
CORE Median. Axile.
CELLS Small. Roundish obovate or obovate.
SEEDS Obtuse, broad, blunt and fairly plump.
LEAVES Medium size. Acute. Bluntly and broadly serrate. Medium thick. Slightly upward-folding some undulating. Mid green. Not downy.
POLLINATION GROUP 4

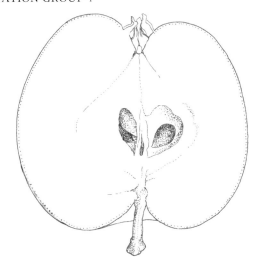

QUEEN

This is a fairly old mid season to late cooking apple. It was raised in 1858 by a farmer named W Bill at Billericay in Essex (UK) apparently from the pips of an apple purchased in the market, and it first fruited in about 1874. The apple was introduced to commerce in 1880 as The Claimant by Messrs Saltmarsh of Chelmsford and was awarded a First Class Certificate by the RHS in November of that year. It is a garden and exhibition cultivar. The trees are moderately vigorous, upright-spreading and partial tip-bearers. The cropping is good, and the picking time is the end of August. The fruits are very acidic, tangy and fruity with a very good flavour and a slight aroma. The flesh is fine-textured, firm and juicy and cooks to a fluff.

SIZE Very large, 89 x 64mm (3$^{1}/2$ x 2$^{1}/2$").
SHAPE Flat. Distinctly flattened at base and apex. Well rounded ribs. Slightly irregular and slightly five crowned at apex. Fairly symmetrical, can be a little lop-sided.
SKIN Very pale whitish-green becoming very pale yellow. Quarter to three quarters flushed with small flecks and dots of greyed-red and because the yellow skin shows through gives the appearance of a rather brownish-orange. Short broken stripes of bright red. Lenticels inconspicuous tiny green or grey dots. Skin smooth and slightly greasy.
STALK Medium thick (3mm). Fairly long (18mm). Usually protrudes beyond base.
CAVITY Deep and very wide. Lined with fine pale ochre russet partly overlaid with scaly grey-brown russet which can streak out over base.
EYE Fairly large, half or completely open. Sepals erect convergent or somewhat divergent. Some stamens showing. Downy.
BASIN Fairly deep. Medium width. Usually definitely ribbed.
TUBE Cone-shaped.
STAMENS Median.
CORE LINE Median.
CORE Median to distant. Axile.
CELLS Ovate or roundish.
SEEDS Acute.
LEAVES Medium to large. Broadly oval. Bluntly serrate. Medium thick. Flat or slightly undulating. Mid green. Very downy underneath.
POLLINATION GROUP 3

NORFOLK ROYAL

This mid season dessert apple appeared as a chance seedling at Wright's Nurseries at North Walsham in Norfolk in 1908. It was introduced in 1928 and named in 1930. It used to be very well known in eastern counties. The trees are moderately vigorous, upright in habit and partial tip-bearers. They are fairly hardy and suitable for colder areas. The season is late September to early December, picking time early September. The cropping is good and the attractive fruits have a pleasant though not strong flavour and a slight, pleasantly sweet aroma. The flesh is creamy white tinged pink near the skin, fairly coarse-textured, moderately firm but not hard, fairly crisp and juicy.

SIZE Medium to large. 70 x 67mm (2³/4 x 2⁵/8″).
SHAPE Conical to long-conical. symmetrical or lop-sided. Slight well rounded ribs. Can be rather flat-sided towards apex. Regular.
SKIN Very pale yellow to pale whitish-yellow. Quarter to almost completely flushed with brilliant red to crimson facing the sun, and flecked dotted and striped with red over the yellow skin on shaded side. Indistinct broad broken stripes of deep brownish-crimson. Lenticels inconspicuous small whitish-ochre or green-ochre dots. Skin smooth and very shiny. Becomes greasy if stored.
STALK Fairly slender (2.5mm). Short to medium (8–15mm). Usually level with base of fruit.
CAVITY Narrow and deep. Lined with varying amounts of grey-brown russet over ochre-green, which can streak out over base.
EYE Fairly small. Closed. Sepals small, sometimes separated at base, erect and slightly convergent with tips reflexed. Very downy.
BASIN Medium width and depth. Ribbed and slightly puckered. Some ochre-brown or grey-brown russet radiating from eye.
TUBE Cone-shaped occasionally funnel-shaped.
STAMENS Basal. Median towards basal when funnel-shaped tube.
CORE LINE Prominent. Almost basal, clasping.
CORE Median to sessile on long conical fruits. Axile.
CELLS Round.
SEEDS Obtuse. Large wide and flat. Blunt. Straight or slightly curved.
LEAVES Medium size. Acute to narrow acute. Crenate or bluntly serrate. Medium thick. Upward-folding. Mid yellowish-green. Some rather downward-hanging. Undersides fairly downy.
POLLINATION GROUP 5

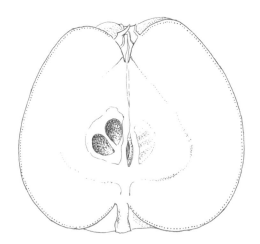

KESWICK CODLIN

This very old second early cooking apple was found, according to Robert Hogg in *The Fruit Manual* 'growing among a quantity of rubbish behind a wall at Gleaston Castle near Ulverstone' in Lancashire. It was introduced by John Sander, a Keswick nurseryman, in about 1790 and he named it Keswick Codlin. It is a typical codlin-type apple, being long and rather angular. The season is late September to October, picking time late August. The trees are prolific bearers though with a biennial tendency. They are moderately vigorous, upright-spreading, spur-bearers and sufficiently hardy to be grown right up into the north of the country. The flesh is yellowish-white, soft and rather coarse-textured and cooks to a froth. The aroma is faintly acid.

SIZE Medium large. 74 x 67mm (2⁷/₈ x 2⁵/₈").
SHAPE Round-conical to oblong-conical. Flattened at base tapering to a flattened apex. Prominent sometimes angular ribs making the fruit flat-sided. Ribs are more pronounced towards apex where they terminate in five distinct crowns. Regular. Symmetrical or a little lop-sided.
SKIN Pale green becoming pale yellow. Can be slightly flushed with pale greyish-orange or darker greyish-brown. Flush often streaked in places. Some have one or more raised russet hair lines running from stalk almost to the apex. Lenticels prominent grey russet dots with pinky-brown surround on the flush, otherwise green surround. Skin smooth and dry becoming slightly greasy.
STALK Stout (4mm) and short (9mm). Often fleshy. Protrudes very slightly beyond base or level with base. Stalk sometimes incorporated within a large fleshy protuberance on side of cavity.
CAVITY Wide and shallow. Regular. Some fine grey-brown russet radiates from stalk.
EYE Medium size. Closed. Rather pinched-in looking. Sepals broad based, long and connivent with tips reflexed. Downy.
BASIN Medium depth and width. Irregular. Often beaded. Some streaky brown russet can radiate out from eye.
TUBE Cone-shaped.
STAMENS Median.
CORE LINE Basal.
CORE Median to rather distant.
CELLS Ovate. Lanceolate.Tufted.
SEEDS Acute. Very plump. Tufted.
LEAVES Small. Acute. Serrate. Medium thick and leathery. Upward-folding and slightly undulating. Mid grey-green. Undersides downy.
POLLINATION GROUP 2

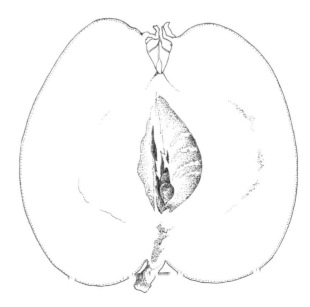

ARTHUR TURNER

This large mid season cooking apple was raised by Charles Turner of Slough, Bucks (UK) and was first exhibited as Turner's Prolific at the RHS in 1912 where it was given an Award of Merit. It was re-named Arthur Turner in 1913 and introduced in 1915. The season is September to November, picking time late September to early October. The blossom received an RHS Award of Garden Merit in 1945. The trees are vigorous, upright and fairly resistant to scab and are suitable for growing in the North. The trees produce spurs very freely and the cropping is regular and good. The fruits have a pleasant mild acidic flavour and require a fairly long cooking time. They break up completely but not to a fluff. The flesh is yellowish-white, coarse textured and rather dry. No aroma.

SIZE Large 80 x 48mm (3^{1}/8 x 1^{7}/8") or 80 x 76mm (3^{1}/8 x 3").

SHAPE Round-conical to oblong. Flattened at base and apex. Usually symmetrical. Slight, well-rounded ribs more pronounced towards apex where it is five-crowned. Irregular.

SKIN Light green much overlaid at base with fine scarf skin. Variably flushed: from delicate blush of greyed-orange to quarter flushed greyed-red. There can be a patch of bright purple-pink at base and small flecks of grey-brown russet. Lenticels inconspicuous pinky-brown russet dots surrounded by white on the green skin. Skin smooth and dry.

STALK Variable. Very stout (4–5mm) and short (10mm) deep within cavity, or medium thick (3mm) and longer (16mm) protruding slightly beyond base.

CAVITY Deep and usually wide. Lined with ochre-brown streaky russet, which can be scaly.

EYE Medium size, partly open. Sepals can be slightly separated at base, rather flat convergent with tips reflexed. Quite downy.

BASIN Quite deep and wide. Slightly puckered. Some brown russet.

TUBE Long funnel-shaped.

STAMENS Median towards marginal.

CORE LINE Median.

CORE Somewhat sessile. Axile.

CELLS Round to slightly obovate. Frequently small.

SEEDS Acute. Fairly wide. Fairly plump. Straight.

LEAVES Large. Broadly oval. Bluntly serrate or crenate. Medium thick. Slightly upward-folding and undulating. Mid grey-green. Slightly downward-hanging. Undersides very downy.

POLLINATION GROUP 3

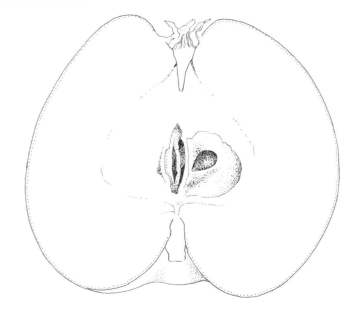

LIMELIGHT

Limelight is a wonderfully crisp and refreshing dessert apple bred by H. Ermen at the East Malling Research Station in Kent (UK) from Discovery and Greensleeves. It was introduced by Mr. Ermen in the 1980s. It makes a heavy cropping, neat, upright-spreading shaped tree of small vigour which is a spur-bearer. It is more frost resistant than some, making it suitable for colder areas, and is very disease free and suitable for wetter areas. The season is September to November, picking time September. The flesh is firm, fine-textured but not hard and quite juicy. It has a delicious flavour, sweet with a good acid balance. Aroma nil. A lovely apple.

SIZE Medium to large 75–80 x 60mm (3″ x 2¹/₂″).
SHAPE Flat-round. Large well rounded ribs. Can be flat-sided. Symmetrical. Fairly regular.
SKIN Pale yellow-green becoming lemon yellow with slightly golden yellow flush. Lenticels conspicuous small grey russet dots. Can be some small flecks of grey russet. Skin smooth and dry.
STALK Short to medium (20–25mm). Fairly slender (2–3mm). Set deep in cavity. Within cavity or protruding slightly beyond.
CAVITY Deep and narrow. Green and mostly lined with fine pale grey-brown russet which can streak out over base with some green streaks.
EYE Quite large. Closed. Sepals connivent with quite long tapering tips reflexed or broken off.
BASIN Medium width. Mostly ribbed and some beading. No russet.
TUBE Funnel -shaped.
STAMENS Median.
CORE LINE Median.
CORE Median.
CELLS Round. Axile open.
SEEDS Acute. Fairly plump.
LEAVES Mid rather dull greyish green. Oval to broadly oval. Bluntly bi-serrate. Flat not undulating. Undersides not downy.
POLLINATION GROUP 4 Reasonably self-fertile, better with a pollinator.

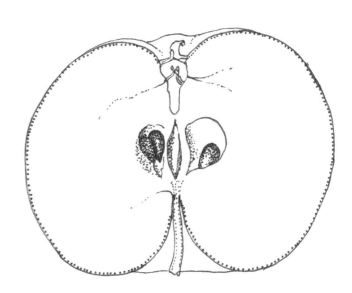

HARVEY

This is one of the oldest English culinary apples. It was mentioned as early as 1629 by John Parkinson, an apothecary who published details of the important fruit varieties available at that time. It was raised in Norfolk and thought to have been named after Dr Gabriel Harvey, Master of Trinity Hall, Cambridge. This variety was well known in East Anglia up to the middle of the last century. The tree is of moderate vigour, upright-spreading and is a partial tip-bearer. The season is late September to January, picking time mid September. The fruits are sub-acid with a good flavour and break up completely when cooked. The flesh is creamy-white, firm, fine-textured and dry. Slight acid aroma.

SIZE Large, 76 x 68mm (3 x 2⅝").

SHAPE Oblong-conical. Base flat, shoulders rounded. Flattened at apex. Broad well-rounded ribs becoming more angular towards apex. Rather flat-sided. Can be irregular. Symmetrical or lop-sided.

SKIN Yellow-green becoming yellow. There can be a slight ochre-brown or greyed-orange flush. Frequently some patchy and spotted scarf skin at base and on cheeks. Lenticels fairly conspicuous pinky-brown or green dots surrounded by scarf skin, becoming smaller and more numerous at apex. Small scuffs and patches of grey-brown russet on cheeks. Variably patched and netted with scaly grey-brown russet at apex and base, usually more on one side than the other. Skin textured and dry.

STALK Short or medium length (12–18mm). Medium thick (3mm). Level with or just beyond base.

CAVITY Narrow and deep. Lined with scaly grey-brown russet which streaks and scatters over base.

EYE Medium to large. Open. Sepals short, convergent or erect convergent, sometimes slightly separated at base. Tips reflexed or broken off.

BASIN Medium width and depth. Slightly ribbed. Some grey-brown slightly scaly russet. Can be beaded. Slightly downy.

TUBE Funnel-shaped.

STAMENS Median.

CORE LINE Median to basal clasping.

CORE Median. Abaxile or axile open.

CELLS Obovate.

SEEDS Acute or acuminate. Broad and fairly plump.

LEAVES Medium to small. Acute. Bluntly serrate or crenate. Medium thick. Flattish. Slightly upward-folding. Light grey-green. Fairly downy.

POLLINATION GROUP 2

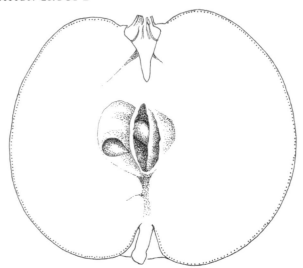

ST EDMUND'S PIPPIN

This early russet was raised in about 1870 by R Harvey at Bury St Edmunds in Suffolk (UK) and is thought to be a chance seedling. It was first recorded in 1875 when it received a First Class Certificate from the RHS. Season late September and October. It is the best early russet for it has excellent flavour, very rich and quite sweet but with a nice balance of acidity, but the fruits bruise easily and it can be unattractive where the russet is thin and patchy.

Picking time mid September. The flesh is creamy-white, tender, juicy and fine-textured. The cropping is uncertain and it can overcrop and bear small fruits and the fruits have a relatively short shelf-life.

SIZE Medium 63 x 52mm (2^1/2 x 2″).
SHAPE Flat-round to round-conical. Well flattened at base and apex. Usually symmetrical but can be lop-sided. Sometimes a hint of one or two well-rounded ribs. Fairly regular.
SKIN Pale greenish-yellow becoming golden-yellow. Some fruits about a quarter flushed bright orange-red. Partly to almost completely covered with fine grey-golden russet overlaid with fine brown scaling. Lenticels inconspicuous as either tiny white dots, dark brown dots or pale ochre dots on the russet, and they are often angular or star-shaped. Skin very dry and slightly rough.
STALK Slender (2mm) and fairly long (12–22mm). Protrudes beyond base.
CAVITY Deep and quite narrow. Lined with russet which circumvents the cavity.
EYE Fairly small. Closed. Sepals short, rather flat-convergent or connivent. Downy.
BASIN Rather narrow. Medium depth. Frequently yellow or green skin showing just around the eye otherwise russet lined. Slightly puckered.
TUBE Cone-shaped.
STAMENS Median or marginal.
CORE Median. Axile.
CORE LINE Median. Rather faint.
CELLS Ovate.
SEEDS Obtuse. Quite large. Blunt. Rather broad and often flattish on one side. Fairly plump and slightly curved.
LEAVES Medium size. Broadly oval. Bluntly serrate. Medium to thin. Can be slightly upward-folding. Not undulating. Light yellowish-green to mid-green. Undersides slightly downy.
POLLINATION GROUP 2

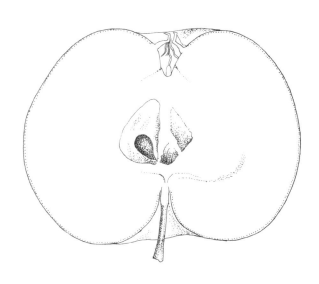

LORD LAMBOURNE

This fairly well known mid season dessert apple was raised by Messrs. Laxton Bros. of Bedford (UK) in 1907 from James Grieve x Worcester Pearmain. It was introduced by Laxtons in 1923. The RHS awarded it the Bunyard Cup in 1921 and an Award of Merit in 1925. The trees are compact in shape, of moderate vigour and are partial tip-bearers. It is a heavy and regular cropper and bears fruit of a uniform size. It is grown mostly in gardens and on a small scale commercially. Season late September to mid November, picking time late September. The fruits are juicy sweet and acidic with a good flavour. The flesh is creamy-white, slightly coarse-textured, firm but tender. Quite strong sweetly scented aroma.

SIZE Medium 64 x 51mm (2¹/₂ x 2").
SHAPE Round slightly conical to flat-round. Usually symmetrical. There can be a trace of well-rounded ribs. Regular.
SKIN Pale greenish-yellow to yellow frequently with a layer of striped green scarf-skin over the top giving it a translucent appearance. Variably flushed: liverish-red with an overlay of green or a fairly bright scarlet. Short broken stripes of darker red. Lenticels distinct pale grey or green russet dots on flush, greenish-brown dots on skin. Skin smooth and slightly greasy becoming more greasy if stored.
STALK Fairly slender (2.5mm) to medium thick (3mm). Medium length (15–20mm). Protrudes beyond base.
CAVITY Medium depth and fairly wide. Lined or partly lined with greenish-ochre and scaly brown russet which can streak out over base.
EYE Small, slightly open. Sepals convergent or slightly connivent with tips reflexed. Very downy.
BASIN Medium width, medium depth to rather shallow. Slightly ribbed, sometimes beaded.
TUBE Cone-shaped.
STAMENS Median to marginal.
CORE LINE Basal, clasping, sometimes meeting.
CORE Median. Axile.
CELLS Round.
SEEDS Acute. Quite large and numerous. Rather broad.
LEAVES Medium size. Acute. Broad and bluntly serrate. Medium thick. Flat not undulating, Upward-folding. Dark grey-green. Undersides slightly downy.
POLLINATION GROUP 2

HERRING'S PIPPIN

This dual-purpose apple was raised by Mr. W.A.Herring of Lincoln (UK) and first recorded in October 1908 when Mr. Herring exhibited specimens at the RHS show. It was introduced commercially by Messrs J.R.Pearson of Lowdham in Nottinghamshire in 1917 and received an RHS Award of Merit in 1920. It is a popular exhibition cultivar. The trees are moderately vigorous, upright-spreading and spur-bearers. Season late September to early November. The cropping is good, picking time early September. It is an attractive apple with fair flavour, sub-acid with a hint of aniseed. The flesh is creamy-white sometimes tinged pink, soft but firm, slightly coarse-textured and not very juicy. It cooks to a dull yellow and stays reasonably intact.

SIZE Large 83 x 76mm (3¹/4 x 3").
SHAPE Round-conical to oblong-conical. Symmetrical or lop-sided. Large broad ribs some of which are fairly well pronounced especially at apex. Five crowned at apex. Slightly irregular.
SKIN Pale greenish-yellow. Half to almost completely covered with red flush, dense on sunny side, more mottled on shaded side. Indistinct broken stripes of deep purplish-crimson. Lenticels inconspicuous, sparse, small ochre or grey-brown russet dots. Much thin streaky scarf skin especially at base spreading on to cheeks. Skin smooth becoming greasy when stored.
STALK Stout (4-5mm). Very short (5-10mm). Sometimes fleshy. Sunk well into cavity.
CAVITY Deep and wide with fine green-ochre and scaly brown russet..
EYE Large, open, with long erect sepals separated at base. Stamens visible. Slightly downy.
BASIN Deep and usually wide. Ribbed. Regular.
TUBE Cone-shaped or slightly funnel-shaped.
STAMENS Median.
CORE LINE Median to basal often running up sides of tube before branching out.
CORE Median. Axile
CELLS Obovate.
SEEDS Obtuse or acute. Fairly large and rather flat.
LEAVES Medium to large. Acute to narrow acute. Serrate. Medium thick, slightly upward-folding and slightly undulating. Mid green. Undersides quite downy.
POLLINATION GROUP 4

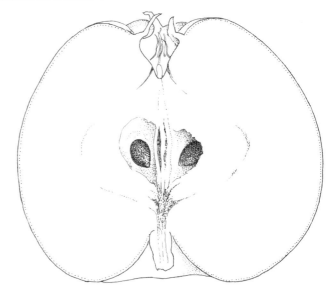

TOM PUTT

A very old variety of small second early to mid season culinary apple, liked by some as a dessert, which used to be popular in the West Country where it was highly esteemed as a cider apple. It was raised in the late 1700s by the Rev. Tom Putt, Rector of Trent in Somerset (UK), reputedly a keen fruit man. Differences in the growth of Tom Putt trees seem to be found probably due to the introduction of local seedlings taken from the original. The trees are vigorous, very spreading and produce spurs freely. The season is September to November. The cropping is good and regular and picking time is early September. The fruits are acidic, with firm greenish-white, coarse-textured flesh and are crisp and juicy. They have a sweetly scented aroma. The skin is rather tough.

SIZE Medium, 65 x 55mm (2$\frac{1}{2}$ x 2$\frac{1}{8}$").
SHAPE Flat-round to round-conical. Rather angular with well pronounced ribs especially noticeable at apex where they terminate in five often knobbly crowns. Irregular. Symmetrical or lop-sided.
SKIN Rather dull yellowish-green to ochre-yellow. Half to three quarters covered with long, sometimes unbroken, broad stripes of greyed-red to red which can run together producing an area of fairly intense stripey red. Lenticels inconspicuous tiny pinky-white or red dots on flush, pale green dots on green skin. Sometimes a hair line present. Skin smooth and almost dry becoming greasy if stored.
STALK Stout (4mm) and rather short (7–15mm). Protrudes beyond base.
CAVITY Fairly shallow to medium. Medium width. Usually some fine downy grey russet. Often a fleshy knob on one side. White or pink lenticels often present within cavity.
EYE Quite large. Closed. Sepals connivent with tips reflexed or broken off. Downy.
BASIN Medium width and depth. Well pronounced ribs. Very puckered and sometimes beaded.
TUBE Cone-shaped.
STAMENS Marginal.
CORE LINE Median.
CORE Median. Axile wide open.
CELLS Ovate.
SEEDS Acute. Fairly large and quite plump.
LEAVES Large. Broadly acute. Serrate. Medium thick, undulating. Dark green. Downy underneath.
POLLINATION GROUP 3

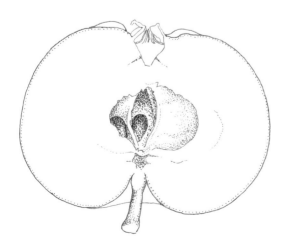

AUTUMN PEARMAIN

This is one of the oldest dessert apples and has been in existence since the late 1500s. It is a second early to mid season apple of reasonable quality. Stocks of this cultivar are somewhat confused. It has been erroneously cultivated as the Royal Pearmain and has been found to be indistinguishable from the Herefordshire Pearmain. The trees are moderately vigorous, upright-spreading and are partial tip-bearers. The cropping is moderately heavy, and picking time is late September. The fruits have a pleasantly sweet though not strong flavour. The flesh is creamy-white tinged green, fine-textured, firm but tender and fairly juicy. They have no aroma.

SIZE Medium, 67 x 60mm (2⁵/8 x 2³/8").
SHAPE Conical. Well rounded shoulders but slightly flattened base. Apex narrow and flattened. Symmetrical or a little lop-sided. Indistinct well-rounded ribs, more pronounced towards apex. Slightly five crowned at apex. Regular.
SKIN Dull green changing to pale yellow. Quarter to three quarters flushed dull gold to greyed-orange. Broken stripes of red to dull reddish-brown away from the sun. Lenticels conspicuous large pale russet dots. Much of the surface netted and flecked with fine grey-ochre or grey-brown russet. Skin slightly textured. Dry becoming greasy if stored.
STALK Fairly stout (3.5mm). Medium length (12–17mm). Extends beyond base of fruit. Usually set at an angle due to lipped cavity.
CAVITY Medium to rather shallow. Fairly narrow. Usually with large lip on one side. Partly or completely lined with fine brown russet which extends thinly over base.
EYE Fairly large half or fully open. Sepals broad based, erect convergent, sometimes flat convergent with tips reflexed, usually separated at base. Very downy
BASIN Medium width. Fairly shallow. Slight trace of ribs. Usually some russet.
TUBE Wide cone or slightly funnel-shaped.
STAMENS Median.
CORE LINE Basal, clasping.
CORE Median. Axile.
CELLS Ovate (Obovate in Hogg's *Fruit Manual*)
SEEDS Accuminate. Plump. Fairly straight.
LEAVES Medium size. Broadly acute. Serrate. Medium thick. Slightly upwarde-folding sometimes undulating. Light yelowish-green. Undersides downy.
POLLINATION GROUP 4

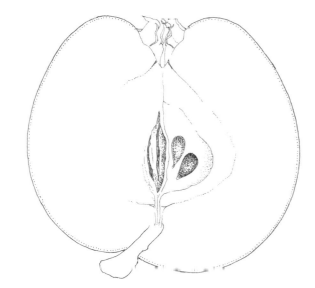

PITMASTON PINEAPPLE

This is a small russetted dessert apple which has a deliciously rich and distinct flavour, nutty, honey sweet with a hint of pineapple. It was grown in England in 1785, bred by a Mr. White who was an employee of Lord Foley of Witley, possibly from a seed of the Golden Pippin. It was introduced by Williams of Pitmaston in Worcester. It has various synonyms, such as Radcliffes Nonpareil, Reinette d'Ananas and Pineapple Pippin.The trees are moderately vigorous, upright in habit and produce spurs quite freely and are noticably scab resistant. The season is mid September to December. Picking time mid September. The flesh is cream, fine-textured and slightly juicy. There is a slight aroma when the fruits are cut.

SIZE Small 50–55mm x 48–50mm (2–2^{1}/8 x 1^{7}/8″).
SHAPE Long-conical to conical. Can be slightly waisted on longer fruits. Slight trace of ribs. A little lop-sided. Flattened at base and apex.
SKIN Greenish-yellow to golden-yellow with a faint flush of deeper golden yellow. No stripes. Usually fairly well covered with fine sometimes rather netted russet with darker flecks of russet in a circular formation round fruit. Hard to distinguish lenticels from flecks of russet. Skin dry and textured.
STALK Quite short (7mm) . Stout (6mm). Often fleshy. Within cavity or protruding slightly beyond.
CAVITY Shallow and narrow. Often some bright green showing. Some fine pale brown russet.
EYE Medium size. Closed or partly open. Often a bit squashed looking. Sepals quite short, well reflexed or broken off, erect convergent.
BASIN Shallow. Pinched looking. Ribbed and often beaded. Mostly green. A bit downy. Can be some tiny amounts of brown russet.
TUBE Cone-shaped.
STAMENS Median or towards basal.
CORE LINE Basal.
CORE Sessile. Axile closed.
CELLS Round.
SEEDS Numerous. Mid brown. Acute. Fairly plump.
LEAVES Small. Quite light blueish green. Acute to broadly acute. Serrate. Fairly thin. Mostly flat and slightly upward-folding. Undersides not downy.
POLLINATION GROUP 3

WEALTHY

An American mid to late season dessert apple raised by Peter M Gideon of Excelsior, Minnesota from seed of the Cherry Crab which he obtained from Albert Emerson at Bangor in about 1860. The cultivar was imported into England during the 1800s and was grown commercially for some while. In 1839 it receivd an Award of Merit from the RHS. The trees are upright-spreading of weak to moderate vigour and partial tip-bearers. The cropping is good and the picking time is mid September. The fruits are crisp, fairly juicy and fresh tasting with a nice tang and are very sweetly aromatic.

SIZE Medum 67 x 61mm (2⁵/8 x 2³/8″) or 67 x 54mm (2 5/8 x 2 1/8″).
SHAPE Round to flat-round. Symmetrical. Some well-rounded ribs. Regular.
SKIN Very pale greenish-yellow becoming whitish-cream. Quarter to nearly completely covered with pinkish-red flush which can be paler and rather striped. Many short broken stripes of crimson often extending into cavity. Whole appearance very striped except on well coloured fruits. Lenticels fairly conspicuous tiny pale ochre dots on flush, less conspicuous pale green dots on yellow skin. Skin smooth, dry and covered with bloom.
STALK Very slender (1.5–2mm).and usually fairly long (18–22mm). Usually extends well beyond base.
CAVITY Deep and narrow. Some slightly scaly ochre-brown russet.
EYE Small. Closed. Sepals connivent with tips reflexed. Not downy.
BASIN Deep and rather narrow. Ribbed and slightly puckered.
TUBE Funnel-shaped or cone-shaped.
STAMENS Median inclined towards marginal. Point of attachment above core line.
CORE LINE. Median, clasping neck of the funnel. Basal clasping with cone-shaped tube.
CORE Distant. Axile
CELLS Ovate.
SEEDS Acute. Large and numerous. Long oval.
LEAVES Medium size. Acute. Serrate. Medium thick. Slightly upward-folding and slightly undulating. Mid yellowish-green. Undersides slightly downy.
POLLINATION GROUP 3

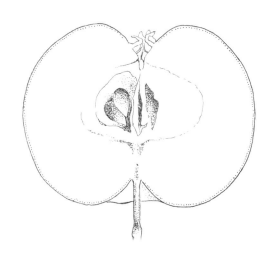

STIRLING CASTLE

This old Scottish cooking apple was raised at Stirling in about 1830 by John Christie, a nurseryman at Causewayhead on the road to the Bridge of Allan. It was introduced by Messrs. Drummond of Stirling and first recorded in 1831. The trees are spreading with weak growth and are spur-bearers. They are hardy and suitable for colder areas. The season is September to December, picking time mid September. The fruit is acid and juicy and I found it full of flavour when eaten as a dessert in September but apparently it improves if left until October. Aroma nil.

SIZE Medium 67 x 57mm (2⁵/8 x 2¹/4") to medium large 76 x 60mm (3 x 2³/8").

SHAPE Flat-round. Flattened at base and apex. Symmetrical. Can be a hint of well-rounded ribs. Regular.

SKIN Bright yellowish-green becoming pale yellow. Some fruits lightly flushed with streaked milky pink or pinky-orange. Varying amounts of streaked or speckled scarf skin chiefly at base. Lenticels conspicuous green or brown russet dots sometimes surrounded by a circle of pink on flush. Skin smooth and shiny becoming greasy.

STALK Variable. Medium (3mm) to stout (4mm). Medium length (12–20mm). Extends beyond base.

CAVITY Deep and wide. Varying amounts of ochre-brown russet which can streak a little over base.

EYE Medium size. Partly open. Stamens fairly broad based some separated at base, erect convergent with tips reflexed but frequently broken off. Slightly downy.

BASIN Deep and wide. Can be a small amount of grey brown russet but usually russet free.

TUBE Cone-shaped.

STAMENS Median.

CORE LINE Median.

CORE Median. Abaxile or axile.

CELLS Roundish, slightly obovate or ovate.

SEEDS Obtuse. Roundish or oval. Blunt. Fairly plump and straight.

LEAVES Medium size. Broadly acute. Serrate. Medium thick. Slightly upward-folding and undulating. Mid green. Undersides downy.

POLLINATION GROUP 3

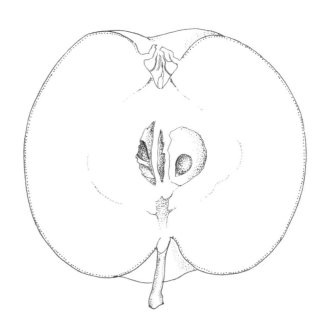

EMPEROR ALEXANDER

This old cooking apple, sometimes listed as Alexander, originated from Russia. It was known in the 1700s but not introduced to England until the early 1800s. It was imported from Riga in Russia by Messrs Lee and Kennedy in 1817, who that year exhibited examples at the London Horticultural Society. It was named as a compliment to the Emperor of Russia to whom apples were said to be sent annually as a present. It is a handsome exhibition variety. The trees are vigorous, upright-spreading, and partial tip-bearers. The season is September to November, picking time mid September. The cropping is moderate. The fruits are sub-acid with no particular flavour and the flesh is white tinged greenish-yellow, coarse-textured, firm and dry. They cook to a pale yellow fluff. Aroma nil.

SIZE Very large. 92 x 76mm (3⁵/8x 3").

SHAPE Round-conical to conical. Very broad at base. Can be five-crowned at apex. Usually symmetrical. Rather angular and flat-sided with slight sometimes angular ribs. Regular.

SKIN Pale yellow-green. Quarter to three-quarters flushed with mottled orange-red. Broad broken stripes and splashes of red. Lenticels inconspicuous ochre russet dots on flush becoming smaller and more numerous pale pink dots towards apex and small grey-brown dots on green skin. Skin smooth and dry becoming greasy.

STALK Fairly stout (3.5mm) and medium length (18mm). Level with base or slightly beyond.

CAVITY Wide and deep. Regular. Lined with scaly light brown or grey russet encircling cavity which can spread slightly over base.

EYE Medium to large. Half open. Sepals erect with tips reflexed. Downy.

BASIN Deep and even. Narrow to medium width. Slightly ribbed.

TUBE Cone or funnel-shaped.

STAMENS Median.

CORE LINE Median or basal.

CORE Median to sessile. Axile. Sometimes open.

CELLS Ovate. Can be tufted.

SEEDS Obtuse. Short and round, Medium size. Fairly plump.

LEAVES Medium size. Broadly acute. Broadly crenate. Medium thick. Slightly upward-folding and undulating. Mid green. Downy.

POLLINATION GROUP 3

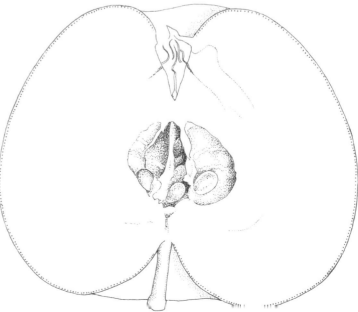

GRAVENSTEIN

There are many theories regarding the origin of this old dual-purpose apple. Some say it was found at Castle Grafenstein, Schleswig-Holstein, in Germany, others that it was sent there from `Italy or Southern Tyrol as 'Ville Blanc' or possibly that scions were sent from Italy by a brother of Count Chr.Ahlefeldt of Graasten Castle, South Jutland. It was thought to have arrived in Denmark about 1669 and in London in 1819. The trees are vigorous, upright-spreading and unsuitable for a small garden. They are spur and tip bearers and triploid. The season is mid September to December, picking time late August and the cropping is fairly good. The fruits are sweet with a good acid balance and distinctive flavour. They make a very nice large dessert apple and also cook to a fine juicy fluff. The flesh is white tinged yellow near the skin, rather coarse-textured, firm, fairly crisp and very juicy. They have a fairly strong sweet and fruity aroma.

SIZE Medium large. 73 x 67mm (2⅞ x 2⅝").to large 83 x 70mm (3¼ x 2¾").
SHAPE Oblong. Large well pronounced ribs running from base to apex. Five-crowned at apex. Can be flat-sided. Often lop-sided. Irregular.
SKIN Yellow-green to greenish-yellow becoming pale yellow. Quarter to three quarters flushed with thin orangey-red which appears in broad bands from base to apex rather than one solid patch. Sparsely striped with greyish-red to crimson. Some scarf-skin at base. Lenticels conspicuous grey-purple or pinky dots on flush, green dots on yellow skin. Skin smooth and dry becoming greasy.
STALK Stout (4mm). Short (10mm). Sometimes a fleshy knob. Within cavity.
CAVITY Deep. Medium width. Partly or fully lined with fine ochre russet which can streak out.
EYE Large. Either closed with sepals connivent, or open with sepals erect and tips reflexed. Very downy which can extend on to basin.
BASIN Medium to wide. Ribbed. Sometimes beaded.
TUBE Funnel-shaped or wide cone-shaped.
STAMENS Median when funnel-shaped, towards basal when cone-shaped.
CORE LINE Indistinct. Median to almost basal.
CORE Somewhat distant. Abaxile.
CELLS Elliptical or round.
SEEDS Acute. Rather small. Sparse. Plump. Curved.
LEAVES Fairly large. Broadly acute. Serrate. Medium to thin. Flat. Dark green. Downy underneath.
POLLINATION GROUP 1. Triploid.

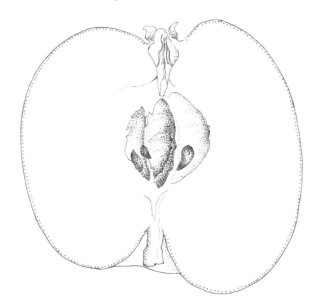

BOUNTIFUL

This is new mid season to late cooking apple produced by East Malling Research Station in Kent (UK) from Cox's Orange Pippin pollinated by Lane's Prince Albert. Bountiful has several advantages over the Bramley in that it is diploid, thus requiring only one pollinator. The trees are of compact habit making them suitable for small gardens and they make cordons easily and can be grown in tubs. It has considerable resistance to apple mildew. The season is September to January and the fruits can be eaten as a dessert apple in late winter. Picking time late September. The cropping is heavy and the fruits have a good flavour though not particularly outstanding. They are sub-acid and require no additional sugar and cook to a yellow fluff. The flesh is creamy-white tinged yellow, firm, fine-textured, tender and fairly juicy. The aroma is sweet but not strong.

SIZE Large 76 x 57mm (3 x 2¼").
SHAPE Round to round conical, flattened at base with slight well rounded ribs and can be rather flat-sided. Symmetrical or slightly lop-sided. Fairly regular.
SKIN Yellowish-green becoming yellow. Can be up to a quarter flushed with brownish-orange, with broken scarlet stripes and flecks. Lenticels inconspicuous purple-grey dots surrounded by very pale yellow on yellow skin and light brownish-orange on flush. Fine grey-brown russet flecking around apex. Skin smooth and dry.
STALK Medium thick (3mm). Medium length (20mm). Protrudes slightly beyond base. Can be set at an angle.
CAVITY Broad and deep. Lined with ochre-green overlaid with light grey-brown russet which can streak out a little over base. Quite pronounced large dots of yellow skin can show through the russet.
EYE Medium size. Partly open with sepals erect and tips reflexed or broken off. Fairly downy.
BASIN Medium width and fairly deep. Slightly ribbed.
TUBE Cone-shaped.
STAMENS Marginal or median.
CORE LINE Basal, clasping.
CORE Median or slightly sessile. Axile, slightly open.
CELLS Obovate.
SEEDS Acute. Regular. Fairly plump.
LEAVES Medium size. Acute. Crenate. Medium thick. Upward-folding. Flat not undulating. Mid grey-green. Undersides very downy.
POLLINATION GROUP 3

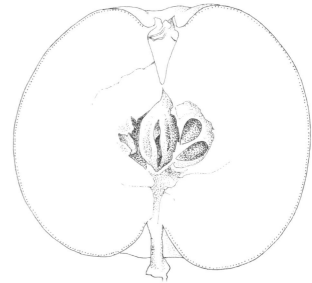

GREENSLEEVES

This fairly new dessert apple was raised in 1966 from James Grieve x Golden Delicious by Dr Alston of East Malling research Station (UK) and received the RHS Award of Merit in 1981. It is a mid season Golden Delicious type and makes a fairly upright compact tree of weakish moderate vigour making it easy to manage. It is hardy and suitable for colder areas and will set a reasonable crop if self pollinated. It is a partial tip-bearer. The cropping is very heavy and the trees bear fruit very early in life. The season is late September to mid November, picking time mid to late September. The fruits are crunchy and sweet with a nice tang in September and October but go rather soft in November. The flavour is good but it soon fades and the skin is a little tough. The aroma is very slight. The flesh is creamy-white, slightly coarse-textured and juicy.

SIZE Medium. 67 x 60mm (2⁵/8 x 2³/8″) or 64 x 57mm (2¹/2 x 2¹/4″).

SHAPE Round to oblong. Rounded at base, flattened at apex. Occasionally lop-sided. Slightly five-crowned at apex with a hint of ribs. Regular.

SKIN Pale green becoming whitish-yellow. There can be a gentle flush of greyed-orange and some very small patches of golden-brown russet. There can be a russetted hair-line. Lenticels conspicuous grey-brown russet dots which are slightly raised giving a surface texture. Skin dry.

STALK Fairly slender (2.5mm). Long (20–22mm). Extends well beyond base. Can be set at an angle.

CAVITY Medium depth and width. Some grey-brown or ochre russet which can spread a little over base.

EYE Medium to quite large. Closed or slightly open. Sepals very long, narrow, erect convergent with tips reflexed. Stamens often visible. Downy.

BASIN Moderately deep and fairly wide. Slightly beaded and slightly ribbed. A smattering of golden-brown russet may be present.

TUBE Funnel-shaped, rather narrow.

STAMENS Median.

CORE LINE Median.

CORE Median. Axile.

CELLS Obovate.

SEEDS Acute. Large. Fairly plump, slightly curved.

LEAVES Medium size. Broadly acute. Bluntly and broadly serrate. Thick and leathery. Upward-folding. Flat. Mid yellowish-green. Undersides not downy.

POLLINATION GROUP 3

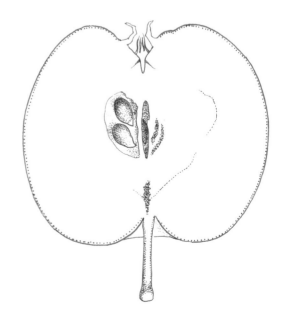

PEASGOOD NONSUCH

This handsome, highly esteemed culinary apple was raised by Mrs. Peasgood of Stamford in Lincolnshire (UK) from a seed, sown in 1858, said to have been from the Catshead Codlin. It was awarded a First Class Certificate by the RHS in 1872. It has always been primarily a garden and exhibition variety being too soft for commercial use. The trees are moderately vigorous, spreading in habit and produce spurs fairly freely. The season is late September to December, picking time mid September. The cropping is heavy with large fruits. The fruits have an extremely good flavour, slightly acid yet quite sweet and they make very good baking apples. The flesh is yellowish tinged slightly green, tender and fairly juicy.

SIZE Large 83 x 70mm (3¼ x 2¾") to very large 92 x 76 (3⅝ x 3").
SHAPE Round slightly flattened to slightly conical. Flattened at base and apex. Symmetrical or a little lop-sided. Regular, not ribbed.
SKIN Pale yellowish-green becoming pale yellow deepening to gold or orange nearest the sun. Quarter to three quarters overlaid with short broken stripes of bright red. There are usually small, slightly scaly, ochre-brown russet patches. Lenticels conspicuous very small ochre-brown, green or pinky dots. Skin smooth, soft and slightly greasy.
STALK Fairly stout to stout (3.5–4mm). Short (8mm).
CAVITY Wide and fairly deep. Green. Some grey-brown russet which can streak out over base.
EYE Medium size. Partly or completely open. Sepals erect or convergent with tips reflexed. Fairly downy.
BASIN Fairly deep and wide. Even. Sometimes slightly puckered.
TUBE Long funnel-shaped, broad at eye.
STAMENS Median.
CORE LINE Median, at junction of bowl and tube of funnel.
CORE Median. Axile.
CELLS Obovate.
SEEDS Acute. Oval pointed, rather tufted. Fairly plump.
LEAVES Fairly large. Broadly oval almost round. Bluntly serrate. Thick and leathery. Mostly flat. Mid fairly bright yellow green. Undersides slightly downy.
POLLINATION GROUP 3

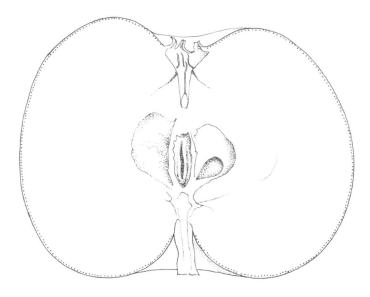

WARNER'S KING

This very old cooking apple was known in the late 1700s as King Apple and under this name Mr Warner, a nurseryman of Gosforth near Leeds, gave it to Mr. Rivers of Sawbridgeworth who re-named it Warner's King. It is said to have originated in an orchard in Weavering Street in Maidstone, Kent. It is a very good heavy cropping variety with a long season, late September to February. The trees are vigorous, upright-spreading and produce spurs very freely. The trees are triploid, fairly hardy and suitable for northern areas but are prone to canker and scab. Picking time mid to end of September. The cropping is heavy with large fruits which have a reasonably good very acid flavour, They cook to a fine fluff. The flesh is white tinged green, rather coarse-textured, crisp and juicy.

SIZE Very large, 95 x 76mm (3³/4 x 3").
SHAPE Flat-round to conical. Broad and flattened at base. Often lop-sided. Well rounded ribs. Irregular.
SKIN Yelowish-green becoming yellow to deeper yellow near the sun. Can be a flush of pinkish or purple-brown. Some small patches and dashes of grey-brown russet. Lenticels numerous and conspicuous greenish-white dots surrounded by dark green or pinkish-brown russet dots. Skin smooth and dry becoming greasy if stored.
STALK Medium thick (3mm). Medium length (17–20mm). Frequently set at an angle.
CAVITY Wide and deep. The ribs can enter cavity. Lined with fine ochre-brown or grey russet which can scatter over base. Often lipped.
EYE Medium size. Closed or partly open. Sepals very long and slender, erect convergent with some tips reflexed. Fairly downy.
BASIN Rather narrow. Medium to deep. Slightly puckered.
TUBE Cone or funnel-shaped.
STAMENS Median sometimes towards basal.
CORE LINE Median sometimes towards basal.
CORE Median. Abaxile.
CELLS Roundish ovate.
SEEDS Acute. Fairly large, angular or curved. Irregular. Often thin.
LEAVES Large. long narrow acute and sharply or bluntly serrate. Fairly thick and leathery. Slightly upward-folding and undulating. Dark green. Downward-hanging. Undersides very downy.
POLLINATION GROUP 2. Triploid.

SATURN

This is a fairly recent introduction from East Malling Research Station in Kent (UK), from PRI 1235 x Starkspur Golden Delicious. It was introduced in 1997. It is a heavy cropping dessert variety, particularly resistant to scab. Although aimed at the organic commercial market, it is a good garden variety and is very suitable for the 'untreated' regime favoured by many. It is self fertile but the cropping is better with a pollinator. The trees are moderately vigorous, upright-spreading and spur-bearers. The season is September to February. Picking time September. The flesh is cream coloured, fine-textured, firm, quite soft though fairly crisp and slightly juicy. The flavour is sweet with a fair acidity. Fruits have no aroma.

SIZE 70 x 65–70mm (2³/4 x 2¹/2–2³/4").
SHAPE Conical to oblong-conical. Pronounced ribs becoming five crowned at apex. Irregular.
SKIN Primrose yellow. Almost completely covered with brilliant scarlet to crimson flush, overlaid with broken crimson to deep crimson stripes extending into the cavity. Lenticels conspicuous on flush as yellow dots with a centre of greeny-grey, and tiny grey dots on yellow skin. Skin smooth and slightly greasy. An attractive apple.
STALK Long (30mm) and slender (2mm). Protruding well beyond cavity. Usually stalk bends to one side.
CAVITY Deep and fairly narrow. Fine greenish ochre russet within cavity.
EYE Small. Closed. Sepals connivent.
BASIN Small. Medium depth to shallow. Ribs visible. Russet free or tiny amount of greenish-ochre russet.
TUBE Funnel-shaped, sometimes widening out again at the base. Usually rather deep.
STAMENS Basal.
CORE LINE Median.
CORE Median to distant.
CELLS Obovate. Axile.
SEEDS Acuminate.
LEAVES Dark rather dull green. Quite large. Medium thick. Fairly flat. Broadly oval. Teeth crenate. Undersides not very downy.
POLLINATION GROUP 3. Self fertile.

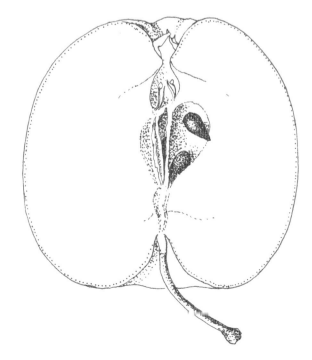

ALKMENE (EARLY WINDSOR)

Alkmene is a dessert apple sometimes known as Early Windsor, and was raised in Germany in the 1930s by Prof. M .Schmidt at the Kaiser Wilhelm Institute, Muncheberg, Brandenburg. The parents were Cox's Orange Pippin and Duchess of Oldenburg. The apple was not introduced until 1972 and received an RHS Award of Garden Merit in 1998. It is widely grown in Germany. Alkmene has an aromatic Cox-like flavour but ripens slightly earlier than the Cox, shows some resistance to scab and mildew and crops regularly and well. The trees are moderately vigorous, upright-spreading and spur-bearers. There are various sports available, such as Red Windsor. The season is October to November, picking time late September. The Fruit has a rich flavour, somewhat juicer than the Cox, with creamy-white fine-textured, firm flesh. The fruits have a sharp aroma when cut, otherwise nil.

SIZE 68 x 62mm (2³/4 x 2¹/2").
SHAPE Conical. Regular. Mostly symmetrical.
SKIN Clear greenish-yellow becoming clear bright yellow to golden yellow. Quarter to three quarters flushed with broken pale orange-scarlet, overlaid with quite long broken red stripes. Lenticels small greenish-grey dots, quite pronounced at apex. Skin smooth and dry, slightly rough at apex.
STALK Mostly short (8–10mm). Medium thick (3mm). Within cavity or protruding slightly beyond.
CAVITY Medium width and depth. Regular. Greenish-grey russet in cavity often spreading over base.
EYE Medium size. Open. Sepals erect. Stamens visible.
BASIN Wide. Medium depth. Ribbed and sometimes puckered. Usually pale green around eye.
TUBE Cone-shaped.
STAMENS Median.
CORE LINE Basal, Clasping.
CORE Median. Axile slightly open.
CELLS Round.
SEEDS Acute.
LEAVES Medium size. Acute to broadly oval. Serrate. Mid green. Undersides downy.
POLLINATION GROUP 2

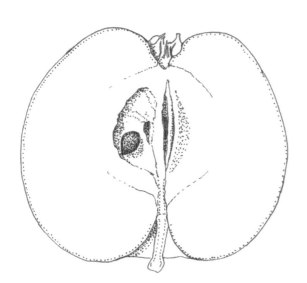

RED DEVIL

This highly striking early dessert apple was raised by H.F.Ermen at Faversham in Kent (UK) in 1975 from Discovery x Kent. It was introduced by Matthews Nursery, Worcester, in 1990. The season is October to November, picking time late September. The flavour is not remarkable unless eaten straight from the tree, when it is sweet, juicy and crisp, tasting slightly of strawberries. Fruits make a pink juice. The trees are upright with good disease resistance and are heavy croppers. The flesh is very coarse-textured, cream tinged red towards skin, with a mild flavour, sweet with a slight acid bite. The skin is a little chewey. Picking time late September.

SIZE Medium 70 x 60mm (2³/4 x 2³/8").
SHAPE Round. Flattened at base. Distinct well rounded ribs unevenly sized often making fruit quite irregular, even squarish. Some trace of ribs at apex. Frequently lop-sided.
SKIN Bright greenish-yellow to bright yellow. Three quarters to completely covered with almost luminous brilliant red and crimson flush. Hardly any trace of stripes. Much covered with conspicuous yellow to ochre yellow dots, some with small dark russet dot in centre, which tend to be larger towards stalk and smaller and more numerous towards eye. Skin smooth and dry. A brilliant and attractive fruit.
STALK Short (7–10mm). Stout (5mm). Often with fleshy knob. Set well within cavity and usually not protruding beyond base.
CAVITY Medium width and shallow. A great deal of scaly russet·over greenish-ochre which spreads out over base. Cavity and shoulder rather uneven, ridged and quite knobbly.
EYE Large. Wide open. Sepals quite long and tapering. Half the sepal reflexed or broken off. Sepals not visible owing to deep tube.
BASIN Medium width. Medium depth. Ribs usually visible. Usually no russet.
TUBE Funnel shaped, often extending into core.
STAMENS Median.
CORE LINE Median towards basal.
CORE Median.
CELLS Round. Axile open.
SEEDS Acute. Mid rich reddish brown. Fairly plump.
LEAVES Broadly acute to broadly oval. Medium size. Mid olive green. Sharply serrate. Medium thick. Flat. Some slightly upward-folding. Undersides slightly downy.
POLLINATION GROUP 3. Self fertile.

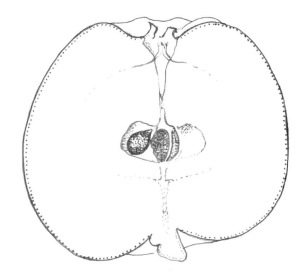

MOTHER (American)

This is an old American mid season dessert apple. It originated at Bolton, Worcester County, Massachusetts and was first recorded in 1844. The *Herefordshire Pomona* records that it was Mr. Rivers of Sawbridgeworth who introduced it to England in the early 1800s. It is noted for its good flavour, sweet and aromatic, but needs full sun to develop its full potential and the cropping is irregular. The trees are moderately vigorous, very upright in habit and produce spurs freely. The season is October to November, picking time late September. It will tolerate a wetter climate. The fruits have a slight sweetly scented aroma and the flesh is creamy-white tinged green towards the core, firm but tender and coarse-textured. The skin is rather tough.

SIZE Medium 64 x 60mm (2^1/$_2$ x 2^3/$_8$").
SHAPE Long-conical. Rounded at base. Rather indistinct broad ribs. Can be flat-sided. Sometimes symmetrical but often lop-sided. Slightly irregular.
SKIN Greeenish-yellow becoming yellow. Quarter to three quarters flushed with dull orange-red to a deeper dull red nearest the sun. Intistinct fairly long narrow stripes of crimson. Some small greyish-ochre russet patches. Lenticels indistinct numerous tiny grey-brown or reddish russet dots. Skin smooth and dry.
STALK Slender (2mm). Medium length (13–17mm) Usually protrudes slightly beyond base.
CAVITY Fairly wide to rather narrow. Medium to fairly deep. Frequently lipped on one side.
EYE Small. Closed or partly open. Sepals erect convergent with tips reflexed. Slightly downy.
BASIN Shallow and rather small. Slightly ribbed and puckered, sometimes a little beaded. There can be a small amount of brown russet present.
TUBE Tiny, funnel-shaped.
STAMENS Almost marginal.
CORE LINE Median towards basal.
CORE Median. Abaxile.
CELLS Elliptical. Tufted.
SEEDS Accuminate or acute. Numerous. Plump.
LEAVES Medium size. Acute. Serrate. Medium thick. Flat not undulating. Slightly upward-folding. Mid grey-green. Undersides slightly downy
POLLINATION GROUP 5

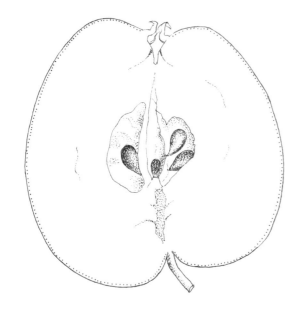

JESTER

Jester is a very pretty dessert apple. It was raised by Dr Alston in 1966 at the East Malling Research Station. The trees are compact, moderately vigorous and produce spurs freely. The cropping is good. The season is October to November, picking time late September. The flesh is fine textured and fairly soft but crisp and juicy. The flavour is fair with a good balance of sugar and acid. Aroma very slight.

SIZE Medium-large, 75 x 65mm (3 x 2½–2¾").
SHAPE Oblong to oblong-conical. Large well-rounded ribs making it occasionally flat-sided. Fairly symmetrical. Sometimes irregular.
SKIN Clear pale greenish-yellow becoming pale yellow. Half covered with bright scarlet flush overlaid with indistinct stripes of slightly deeper red. Lenticels numerous small olive-green dots noticeable on yellow skin. Skin smooth and greasy.
STALK Fairly short (12mm). Slender (2mm). Usually within cavity or protruding slightly beyond.
CAVITY Deep and narrow. Greenish-ochre. Russet lining cavity radiating out slightly. Some brown scaling within.
EYE Medium. Open. Sepals erect, separated at base.
BASIN Fairly deep and narrow. Ribbed and somewhat pinched looking. Some bright green around sepals. Fairly downy.
TUBE Cone-shaped.
STAMENS Basal.
CORE LINE Basal.
CORE Distant. Axile open.
CELLS Round.
SEEDS Acute to acuminate.
LEAVES Acute. Dull mid green. Rather bluntly serrate. Medium thick. Flat. Slightly upward-folding. Undersides not very downy.
POLLINATION GROUP 4

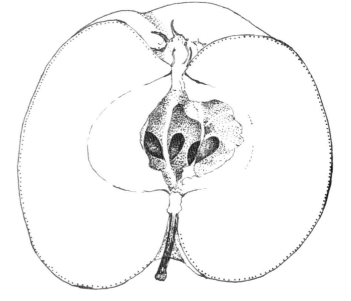

EGREMONT RUSSET

The origin of this apple appears to have been lost. It is thought to have originated in England and was first recorded here in 1872. It has been cataloged by most nurseries since the early part of the last century and is an important apple commercially. It received the Award of Merit from the RHS in 1980. The young tree crops regularly and produces a good sized fruit but the mature tree tends to be biennial in cropping. The trees have moderate vigour, are upright in habit and produce spurs extremely freely. They will tolerate colder and wetter parts of the country. The season is October to December, picking time late September. The fruits are sweet with a rich nutty flavour but the skin is rather tough. The flesh is cream tinged yellow, crisp and firm, fine-textured and fairly dry. It has no aroma.

SIZE Medium 64 x 48mm ($2^1/2$ x $1^7/8$″) to 67 x 57mm ($2^5/8$ x $2^1/4$″).

SHAPE Flat-round, Well flattened at base and apex. Symmetrical or lop-sided. Usually no ribs but can have a hint of one.

SKIN Yellowish-green becoming golden yellow. Often up to half flushed with brownish-ochre to light brownish-red, Half or more covered with fine ochre to grey-brown russet. Lenticels very conspicuous whitish dots becoming larger towards base. Skin very dry.

STALK Medium thick (3mm) and very short (6–10mm). sunk well down within cavity.

CAVITY Medium depth to rather shallow. Narrow. Frequently greenish-yellow with russet thinly spread and skin beneath showing through, particularly at shoulder.

EYE Quite large, usually wide open. Sepals erect or slightly convergent with tips reflexed, exposing stamens.

BASIN Medium width and depth. Can be slightly ribbed or puckered. Regular.

TUBE Funnel-shaped.

STAMENS Median towards basal.

CORE LINE Almost basal.

CORE Median. Axile.

CELLS Small. Roundish oval to oval.

SEEDS Obtuse. Oval, very blunt.

LEAVES Medium size. Acute. Sharply and broadly serrate. Flat not undulating sometimes slightly upward-folding. Thin. Mid green. Slightly downy underneath.

POLLINATION GROUP 2

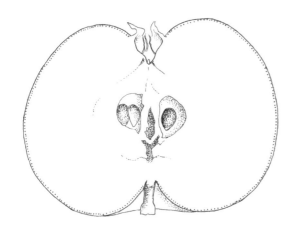

KING OF THE PIPPINS

The origin of this old dessert apple seems somewhat confused.It is thought to be of English origin and introduced in the early 1800s by Mr. Kirke, a nurseryman of Brompton, who named it King of the Pippins. An older name for this apple was Golden Winter Pearmain. It was planted extensively in the past for commercial use. The trees are moderately vigorous, upright in habit and produce spurs very freely. They are reasonably hardy and will tolerate a wetter climate. The season is mid October to December, picking time early October. The fruits are sweet, crisp and juicy with a very rich and vinous, rather nutty flavour. The flesh is creamy-white, firm, fine-textured and crisp. A distinctive and lovely apple.

SIZE Medium 67 x 64mm (2⅝ x 2½") to 60 x 57mm (2⅜ x 2¼").
SHAPE Oblong-conical. Frequently a little lop-sided. Slight trace of well-rounded ribs more evident at apex where it may be five-crowned. Regular.
SKIN Greenish yellow becoming yellow with some green stripes or patches. Quarter to three quarters flushed with brownish-orange with short broken stripes and an occasional long stripe of bright red. Some small pale grey russet flecks and patches. Lenticels conspicuous green or pale ochre russet dots on yellow skin and whitish russet dots on flush. Skin smooth and almost dry.
STALK Variable. Either fairly slender (2.5mm) and long (22mm) or stout (4mm) and short (10mm). Usually extends beyond base.
CAVITY Medium width and depth. Lined with fine greenish-ochre russet which can streak and scatter over base. Cavity bright green where not covered with russet.
EYE Large. Partly open. Sepals broad based and very long. Connivent with tips well reflexed. Fairly downy.
BASIN Rather shallow and wide. Slightly ribbed.
TUBE Funnel-shaped
STAMENS Median
CORE LINE Basal, clasping.
CORE Median. Axile open.
CELLS Obovate.
SEEDS Obtuse or acute. Round and plump.
LEAVES Medium to small. Oval. Serrate. Medium thick. Flat. Mid blue-green. Downy underneath.
POLLINATION GROUP 5 Biennial.

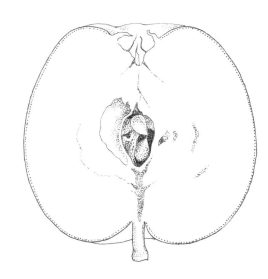

CHARLES ROSS

This handsome dual purpose apple was raised from Peasgood Nonsuch x Cox's Orange Pippin by Charles Ross, gardener to Capt. Carstairs at Welford Park in Derbyshire (UK) from 1860–1908. It was originally named after Thomas Andrew Knight, the President of the RHS. It was first exhibited in 1890, received the Award of Merit in 1899 and in that year, at Capt. Carstair's request, the name was changed and the apple received a First Class Certificate as Charles Ross. The trees are moderately vigorous, upright-spreading and produce spurs very freely. They are fairly resistant to scab but prone to capsid bug and suitable for colder and wetter areas. The season is October to December, picking time mid September. The fruits are juicy, sweet and flavoursome but become flavourless by late October. The flesh is creamy-white, rather coarse, crisp and juicy, It becomes yellow when cooked, sweet, with reasonable flavour and stays fairly intact. Fruits are sweetly aromatic.

SIZE Large 80 x 70mm (3^{1}/$_8$ x 2^{3}/$_4$").
SHAPE Round, slightly conical. Flattened at base and apex. Usually symmetrical but can be lop-sided. No ribbing. Regular. A heavy apple.
SKIN Greenish-yellow. Half to three quarters flushed orange-red to ochre-orange away from the sun. Broad broken stripes of red often overlaid with grey scarf skin at base and on cheeks. Lenticels conspicuous ochre russet dots on flush, grey-brown russet dots on yellow skin. They can be angular or star-shaped. Some russet dots and patches especially at base. Numerous whitish dots at apex. Skin smooth becoming greasy.
STALK Stout to very stout (4–6mm). Short (10mm). Often fleshy. Usually sunk well within cavity.
CAVITY Wide. Medium depth. Usually green. Lined with grey or brown slightly scaly russet. Scarf skin can be in cavity. Sometimes lipped.
EYE Medium size. Open or partly open. Sepals erect convergent with tips reflexed. Very downy.
BASIN Medium depth to shallow. Fairly wide. Even. Can be ribbed.
TUBE Funnel-shaped, rather long and narrow.
STAMENS Median or towards marginal.
CORE LINE Median joining at funnel neck.
CORE Median. Axile.
CELLS Roundish obovate. Rather small.
SEEDS Obtuse. Fairly plump. Regular and straight.
LEAVES Medium size. Acute. Crenate. Medium thick. Slightly upward-folding and undulating. Mid yellow-green. Slightly downy underneath.
POLLINATION GROUP 3

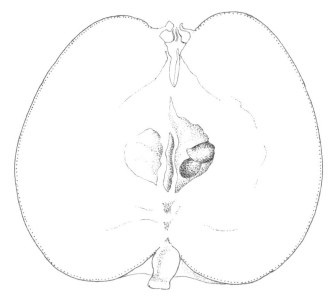

COX'S POMONA

This apple was raised in England in about 1825 by Richard Cox of Colnbrook Lawn, Slough, Buckinghamshire. It is a mid to late season culinary or possibly dual purpose apple, said to be a seedling of Ribston Pippin possibly crossed with Blenheim Orange. It was introduced by Mr. Smale of Colnbrook Nursery in Slough. The trees are moderately vigorous, upright-spreading, spur-bearers. The season is October to December, picking time early September. The cropping is good. As a dessert apple I found it crisp, juicy and flavoursome. When cooked it has a fine, delicate flavour, sub-acid, not requiring additional sugar and it stays fairly intact. The flesh is white tinged slightly green, rather coarse-textured, firm but tender. The aroma is slight but sweetly fragrant.

SIZE Large. 83 x 64mm (3^1/4 x 2^1/2").
SHAPE. Flat-round to round-conical. Well flattened at base and apex. Well pronounced ribs. Five crowned. Symmetrical or lop-sided. Irregular.
SKIN Pale yellow-green to pale greenish-yellow. Quarter to three quarters flushed brilliant red with crimson stripes, fading at the edges to paler red where there are prominent broad stripes and speckles of the brighter red. Indistinct small pinky-white lenticels on flush, more numerous and elongated at base. A brilliant and attractive apple. Skin smooth and slightly greasy becoming more greasy if stored. Very shiny.
STALK Variable. Stout (4mm) and short (10mm) or fairly slender (2.5mm) and longer (15–20mm). Very deeply sunk into cavity.
CAVITY Deep and wide. Usually green or yellow and sometimes lined with grey-brown russet.
EYE Medium. Open. Sepals short, separated at base, erect with tips often broken off. Fairly downy.
BASIN. Deep and rather narrow. Irregular. Ribbed.
TUBE Deep cone-shaped.
STAMENS Median. Axile.
CORE LINE Median, appearing to pull the sides of the tube apart.
CORE Median.
CELLS Roundish-obovate.
SEEDS Obtuse. Rather broad. Fairly plump.
LEAVES Medium size. Acute. Crenate. Medium thick. Slightly upward-folding and very slightly undulating. Dark grey-green. Very downy underneath.
POLLINATION GROUP 4

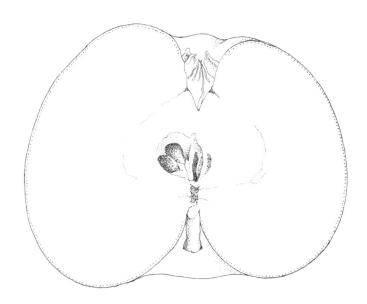

LORD DERBY

This well known mid to late season culinary apple was raised in England by Mr. Witham, a nurseryman of Stockport in Cheshire. It was first recorded in 1862. It is a good variety for marketing before Bramley's Seedling but should be marketed while the fruits are green before they turn yellow. It is a prolific and regular cropper but may require thinning to obtain large fruits. The season is October to December, picking time late September. The trees are moderately vigorous, upright-spreading, show a good resistance to scab and suceed well on wet soils. They are very hardy and suitable for a colder climate. The trees produce spurs freely and the flowers show a good degree of self-fertility. The fruits are light in weight, sub-acid with good flavour and stay intact when cooked. The flesh is greenish-white, slightly coarse-textured, soft and rather dry. They have a pleasant slightly acid but not strong aroma.

SIZE Large 83 x 70mm (3¹/4 x 2³/4").
SHAPE Round-conical to oblong-conical. Pronounced rather angular ribs. Often flat-sided or slab-shaped. Flattened at base, five-crowned at apex. Can be lop-sided. Irregular.
SKIN Strong, bright grass-green becoming bright yellow. Much overlaid wth scarf skin that starts at the base and extends partly up the cheeks making the skin colour rather milky green. Occasionally a slight blush of pinkish-brown nearest the sun. Lenticels fairly conspicuous pale green dots, smaller and more numerous towards apex. Skin smooth and dry.
STALK Very stout (6mm) often fleshy. Very short (5–6mm). Set within cavity.
CAVITY Wide and fairly deep. Green with some streaky scaf-skin. Occasional tiny patch of fine pink-brown russet next to stalk.
EYE Medium size. Closed or partly open. Sepals broad-based, connivent with tips often reflexed right back. Downy.
BASIN Deep. Angular, often pinched-looking. Much ribbed and puckered. Sometimes beaded.
TUBE Deep cone-shaped or funnel-shaped.
STAMENS Median towards marginal
CORE LINE Median.
CORE Median. Abaxile.
CELLS Roundish ovate, or rather elliptical.
SEEDS Obtuse to acute. Regular not curved.
LEAVES Medium size. Acute. Serrate. Medium to thin. Upward-folding very slightly undulating. Mid blue-green. Undersides very downy.
POLLINATION GROUP 4

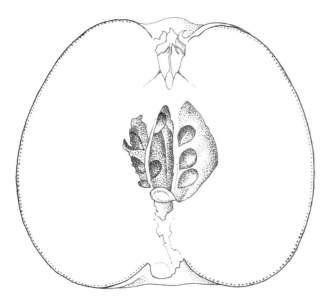

McINTOSH RED

This mid to late dessert apple is of Canadian origin. It was discovered in 1796 by John McIntosh at Matilda Township, Dundas County, Ontario, said to be a chance seedling. It was propagated by Allan McIntosh and named in about 1870. It is widely grown in Canada and the United States. There is a race of McIntosh-type apples, including Tydeman's Early Worcester and Vista Bell, which are noted for their highly perfumed smell and flavour, and there are many highly coloured clones in existance. The trees are moderately vigorous, spreading, and produce spurs very freely. They are however susceptible to canker. The season is October to December, picking time mid September. The cropping is good and the fruits are sweet with a strong vinous flavour. The flesh is very white, tinged pink near the skin, fine-textured, tender almost soft and very juicy. The aroma is strong and very sweetly perfumed.

SIZE Medium large, 70 x 67mm (2³/4 x 2⁵/8").
SHAPE Round to flat-round with fairly well defined ribs which are more noticeable towards apex. Can be five-crowned at apex. Symmetrical or a little lop-sided. Irregular.
SKIN Yellow-green becoming whitish yellow. Half to almost entirely flushed with crimson to brownish-crimson on well coloured fruit. Short broken purple-crimson stripes which are lighter but more noticeable on paler flush. The fruit is covered with a fine lilac bloom which rubs off when touched and the fruit polishes to a high shine. Lenticels fairly noticeable small yellow or white dots. Skin smooth and dry.
STALK Slender to medium (2–3mm). Fairly short (10–20mm). Level with base or slightly beyond.
CAVITY Medium width and fairly deep. Can be slightly ribbed. Partly lined with green-ochre or slightly scaly brown russet.
EYE Very small. Tightly closed or slightly open. Sepals small, erect convergent. Very downy.
BASIN Rather small. Medium depth and rather narrow. Slightly ribbed with some beading.
TUBE Conical
STAMENS Median.
CORE LINE Basal, clasping or towards median.
CORE Median. Abaxile.
CELLS Round.
SEEDS Fairly large. Acute. Plump sometimes angular.
LEAVES Medium size. Acute. Bluntly serrate or crenate. Medium thick. Flat. Slightly upward-folding. Light yellow-green. Very downy.
POLLINATION GROUP 2

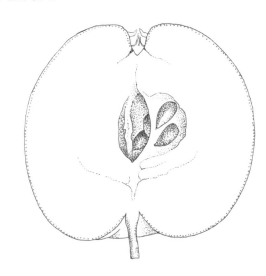

INGRID MARIE

An attractive dessert apple, more popular in northern Europe than in the UK, It was found in the garden of Hoed School near Flemloese on the island of Fyn in Denmark in about 1910, named after a teacher Mr Madsen's daughter who died very young. It was long thought to be a chance seedling of Cox's Orange Pippin. In 2003 it was confirmed to be a hybrid of Cox and the Danish cultivar Guldborg named after a village on the island of Lolland. It is an attractive apple of only fair quality. In October I found the flesh to be crisp and juicy, fine textured with a reasonable if slightly acid flavour but it quickly deteriorates so is best eaten straight from the tree. Ingrid Marie is the parent of Elstar. The trees have moderate vigour, are upright-spreading and spur-bearers. They are resistant to Scab and Mildew and grow well in a warm humid climate. The season is October to December. The picking time is late September. The cropping is good. The flesh is cream tinged green with the red skin colour bleeding into the flesh. Fruits have a slight aroma.

SIZE Medium 75 x 60mm (3 x 2¼").
SHAPE Flat-round. Well flattened at base and apex. No ribs or very slight trace of a well rounded one. Mostly regular. Frequently lop-sided.
SKIN Fairly dull greenish-yellow. Half to almost completely covered with mottled dull scarlet on shaded side to deep crimson on well coloured fruits. Slight trace of broken deeper crimson to purple stripes. Much covered with scruffy russet scuffs and patches and frequently large yellow dots containing a lenticel or star shaped russet mark.Skin smooth and slightly greasy.
STALK Medium length (15mm). Medium width (3–4mm). Protrudes beyond base.
CAVITY Medium depth. Fairly wide. Green with pale brown scaly russet.
EYE Quite large. Open. Sepals separated at base. Tips well reflexed or broken off. Stamens visible.
BASIN Fairly wide. Shallow. Slight trace of ribs.
TUBE Cone-shaped.
STAMENS Median.
CORE LINE Median.
CORE Median or slightly sessile.
CELLS Round. Axile closed.
SEEDS Mid brown. Not very plump.
LEAVES Fairly bright mid to dark green. Medium to small. Acute. Crenate. Fairly flat. Slightly upward-folding. Fairly thin. Undersides downy.
POLLINATION GROUP 4. Self fertile.

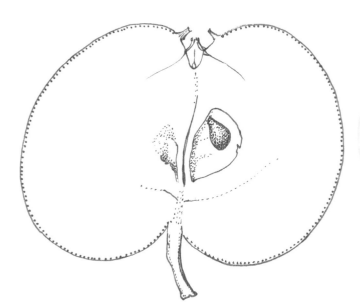

ROSS NONPAREIL

This old high quality dessert apple originated in Ireland. There is a record of it in Meath in 1902. It was introduced to England in 1819. It is often called an Irish russet apple but it is thought to be of French origin. It was thought lost to cultivation but was rediscovered by the RHS after World War 11. It is primarily a garden apple being too small for commercial use. The trees are of moderate vigour, upright-spreading and spur-bearers. The cropping is good and trees are very scab resistant making them suitable for wetter areas. The season is October to December, picking time late September. The flesh is greenish-white, fine-textured and rather dry with rich aromatic flavour, sweet with a good acid balance. Aroma nil, slightly woody when cut.

SIZE Medium 60–65mm x 52–54mm (2³/8–2¹/2 x 2–2¹/8").
SHAPE Round conical to round. Can be very slight trace of one or two ribs but mostly no ribs. Flattened at apex, broadest at base. Regular. Can be slightly lop-sided.
SKIN Very pale green to greenish-yellow. Hardly up to half covered with dull orange-red flush with visible dark red stripes. Half to almost completely covered with fine russet, varying in colour from ochre and greenish-ochre to pale grey-brown. Stripes often visible beneath the grey-brown giving it a pale milky red look–the colour of a ripening strawberry. Russet often broken and patchy over flush giving apple a rather scruffy appearance. A pretty apple in the sunshine.Skin dry and slightly textured.
STALK Medium to long. 20–25mm. Fairly slender (3mm).
CAVITY Narrow to medium. Deep. Often with fleshy swelling on one side. Russetted with ochre russet and overlaid with fine scaly grey-brown russet.
EYE Medium size. Open. Sepals erect convergent with long tips well reflexed or broken off. Stamens visible.
BASIN Fairly narrow and fairly shallow. Mostly regular. No ribs. Lightly russetted with fine light brown russet.
TUBE Cone-shaped.
STAMENS Marginal.
CORE LINE Basal.
CORE Median.
CELLS Round. Axile closed.
SEEDS Fairly light brown. Acute. Plump.
LEAVES Broadly oval. Small. Serrate with very small serrations. Very slightly upward-folding. Slightly undulating. Undersides downy.
POLLINATION GROUP 2

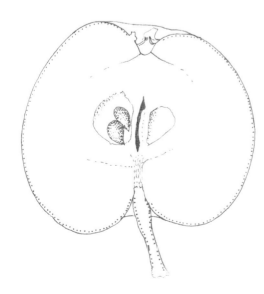

ROYAL JUBILEE

A mid season culinary apple which was raised by John Graham of Hounslow in Middlesex (UK). The parentage is unknown. It was first recorded in 1888 and introduced by G Bunyard & Co. of Maidstone, Kent in 1893. The trees are weak, upright-spreading and spur-bearers. The season is October to December, picking time late September. The cropping is good and regular though the fruits tend to drop all at once so it needs watching at harvest time. It is late flowering which makes it useful for growing in areas subject to late frosts. The fruits are sub-acid with a pleasant but weak flavour and remain intact when cooked. They are slightly aromatic. The flesh is creamy-yellow, firm, coarse-textured and juicy.

SIZE Medium 67 x 67mm ($2^5/8$ x $2^5/8$") or 70 x 54mm ($2^3/4$ x $2^1/8$") or large 83 x 73mm ($3^1/4$ x $2^7/8$").

SHAPE Oblong to conical. Very irregular with large ribs, some more prominent than others. Can be flat-sided, especially towards apex. Usually lop-sided.

SKIN Greenish-yellow becoming yellow with deeper yellow nearest the sun. There can be a slight flush of pale ochre-brown deepening to pinkish-brown facing the sun. There can be some patchy scarf skin at or towards base. Lenticels conspicuous green or grey-brown russet dots, or white dots surrounded by a brown circle on the brighter flush. Some fruits have a light dusting of grey-brown russet. Skin smooth and slightly greasy becoming more greasy if stored.

STALK Stout (4mm). Short (10mm). Within cavity or level with base. Sometimes fleshy.

CAVITY Fairly wide. medium depth. Partly lined with scaly golden russet.

EYE Medium size. Closed or slightly open. Sepals broad based, erect and connivent. Tips reflexed or broken off. Moderately downy.

BASIN Medium depth and fairly narrow. Extremely pinched.Can be some large fleshy beads.

TUBE Funnel-shaped. Often extending into core.

STAMENS Median.

CORE LINE Median.

CORE Median to distant. Abaxile.

CELLS Obovate or elliptical.

SEEDS Acuminate. Plump

LEAVES Medium size. Acute. Bluntly pointed. Rather thin. Upward-folding, occasionally downward-folding. Mid yellowish-green. Downward-hanging. Undersides fairly downy.

POLLINATION GROUP 5

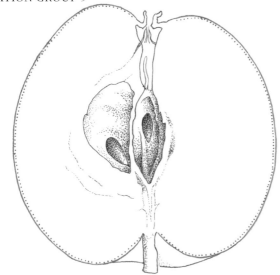

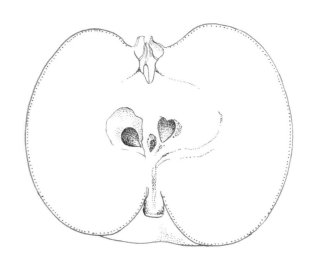

RIVAL

A well known mid to late dual purpose apple which was raised in England by Mr. Charles Ross of Newbury in Berkshire from Peasgood Nonsuch x Cox's Orange Pippin. It was introduced by Clibrans of Altrincham and first recorded in 1900 when it received the Award of Merit from the RHS. The trees are moderately vigorous, spreading in habit and spur-bearers. They are fairly hardy and also suitable for wetter areas. The season is October to December, picking time late September to early October. The cropping is good though it tends to be biennial. The fruits are slightly acid with a reasonably good flavour. It is a very refreshing crunchy apple when eaten as a dessert and has a nice tang. The flesh is creamy-white, very crisp and juicy and fine-textured.

SIZE Medium large 73 x 58mm (2⁷/₈ x 2¹/₄").
SHAPE Flat-round. Well flattened at base and apex with slight trace of well rounded ribs. Symmetrical or slightly lop-sided. Fairly regular.
SKIN Pale yellowish-green. Quarter to half flushed with bright but thin orangey-red with fairly distinct short broken stripes of bright red. A delicate network of fine grey-ochre russet present in varying amounts. Lenticels inconspicuous as greenish-white or grey-brown dots. Skin smooth and dry.
STALK Medium to stout (3–4.5mm). Short (10mm). Sunk well within cavity.
CAVITY Fairly wide and deep. Lined with fine ochre or green-brown russet which can scatter over base.
EYE Medium size. Partly or wide open. Sepals short, erect, slightly separate at base when eye is wide open, with tips reflexed. Stamens usually present. Fairly downy.
BASIN Wide and deep. Slightly ribbed. Small amounts of fine brown russet can be present.
TUBE Funnel-shaped.
STAMENS Median.
CORE LINE Median.
CORE Median. Axile.
CELLS Round.
SEEDS Acute. Roundish and pointed. Fairly plump and straight.
LEAVES Medium size. Oval. Bluntly serrate. Medium thick. Slightly upward-folding. Not undulating. Mid green. Undersides downy.
POLLINATION GROUP 3. Biennial.

NORFOLK BEAUTY

A mid season culinary apple of English origin that was raised by Mr. Allan at Gunton Park in Norwich, thought to be from Warner's King x Waltham Abbey. It was first recorded in 1901 when it received an RHS Award of Merit, and was introduced in 1902 when it received a First Class Certificate. The trees are reasonably hardy and well known in Eastern counties, as the name suggests. They make vigorous growth, are spreading in habit and are partial tip-bearers. The season is October to December, picking time early September. The cropping is good and the flavour of the fruit is reasonable. When cooked it is pale cream in colour, acidic, and breaks up completely though not to a fluff. The flesh is creamy-white, coarse-textured and fairly juicy. The fruits have no aroma.

SIZE Large 80 x 67mm ($3^1/8$ x $2^5/8$").
SHAPE Round to round-conical. Flattened at base. Symmetrical or lop-sided. Usually fairly distinct broad ribs which can be angular, with one sometimes more pronounced making it irregular, otherwise fairly regular. Can be flat-sided.
SKIN Pale yellow-green becoming pale whitish-yellow. There can be a slight flush of pinkish-brown. Lenticels conspicuous white dots surrounded by an areola of green, or indistinct pinkish brown dots, becoming smaller and more numerous towards apex. Skin smooth and dry.
STALK Medium (3mm). Short (8–10mm). Usually level with base.
CAVITY Narrow. Medium to shallow in depth. Lined with scaly brown russet which can streak out over base.
EYE Large. Open. Sepals erect convergent or almost flat convergent, separated at base. Fairly downy. Stamens often visible.
BASIN Medium depth and width. Rather irregular being slightly ribbed and puckered.
TUBE Wide and shallow. Cone-shaped.
STAMENS Median.
CORE LINE Basal clasping. Median if tube is funnel-shaped.
CORE Median. Abaxile.
CELLS Eliptical. Tufted.
SEEDS Usually acuminate, some acute.
LEAVES Medium to large. Broadly acute. Finely serrate. Rather thin. Slightly upward-folding and undulating. Mid blue-green. Fairly downy underneath.
POLLINATION GROUP 2

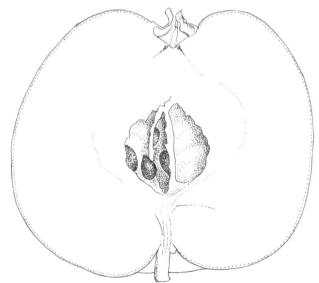

ALLINGTON PIPPIN

This mid to late dessert apple was raised in Lincolnshire (UK) by Thomas Laxton some time before 1884. It received an RHS First Class Certificate in 1889 when exhibited by W & J Brown of Stamford as Brown's South Lincoln Beauty and in 1894 it was re-named Allington Pippin and received the Award of Merit. It was introduced by G Bunyard & Co in 1896. Earlier last century it was widely grown in Kent, Cambridgeshire and Ely. The trees are fairly hardy and suitable for the North and West. They are moderately vigorous, upright-spreading and spur-bearers and make a lot of young growth. The cropping is heavy but tends to be biennial. The fruits are dull looking but have a rich aromatic flavour and are crunchy yet not hard. They are often confused with Laxton's Superb. The flesh is creamy-white, fine-textured and juicy with no aroma.

SIZE Medium large 70 x 61mm (2³/4 x 2³/8").
SHAPE Conical. Trace of well-rounded ribs. Lop-sided or symmetrical. Regular.
SKIN Pale whitish-green to lemon yellow becoming pale yellow. Quarter to three quarters flushed with greyed-red or greyed-orange. Indistinct broken stripes of deeper red. Much overlaid with scarf skin giving it a bloomed appearance. Lenticels conspicuous grey-brown russet dots sometimes star-shaped. Some grey-brown or ochre russet patches. Skin smooth and dry.
STALK Medium to fairly stout (3–3.5mm). Fairly short (15mm). Protrudes slightly beyond base or level.
CAVITY Medium width. Medium to fairly deep. Regular. Some grey-green scaly russet streaks and scatters out over shoulder.
EYE Medium size partly or fully open. Sepals very long and tapering, erect or convergent with tips well reflexed. Not downy.
BASIN Medium depth. Fairly wide. Slightly ribbed. Occasional patch of grey-brown russet.
TUBE Funnel-shaped.
STAMENS Basal.
CORE LINE Basal, clasping.
CORE Median. Axile closed or open.
CELLS Ovate.
SEEDS Numerous. Acute.
LEAVES Medium to small. Acute. Crenate or bluntly serrate. Medium thick. Slightly upward-folding. Flat. Yellow-green. Undersides downy.
POLLINATION GROUP 3. Biennial.

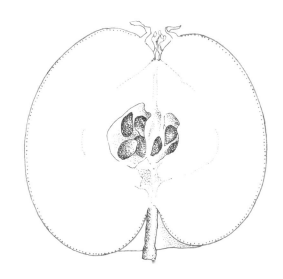

SUNSET

A high quality mid to late season dessert apple which was raised by Mr. G.C Addy at Ightham in Kent (UK) from a pip of Cox's Orange Pippin and introduced jointly with Mr William Rogers of Dartford in Kent. It was named in 1933 and received an RHS Award of Merit and a First Class Certificate in 1982. It is a good garden variety being too small for commercial use and can be grown in more moist areas where Cox's Orange Pippin may fail. Trees are moderately vigorous, reasonably hardy, upright-spreading and compact. They produce spurs very freely. The cropping is good and reliable though fruits need thinning to achieve a decent size. The season is October to December, picking time late September. The flesh is creamy-white, firm, slightly coarse-textured and fairly juicy with a deliciously robust flavour.

SIZE Medium 61 x 51 (2³/8 x 2").
SHAPE Flat round to round-conical. Symmetrical or lop-sided. Trace of well rounded ribs. Fairly regular.
SKIN Greenish-yellow becoming yellow. Quarter to three quarters flushed with bright orange to scarlet. Distinct broad broken stripes of crimson. Numerous small patches of fine ochre to grey-brown russet. Lenticels fairly conspicuous large grey-brown russet dots sometimes angular or star-shaped. Skin smooth, dry and slightly textured.
STALK Medium to stout (3–4.5mm). Medium to fairly long (15–22mm). Extends well beyond base.
CAVITY Wide. Medium to fairly deep. Regular. Partly or more lined with fine ochre or scaly grey-brown russet which can scatter over base.
EYE Medium size. Half open. Sepals erect convergent with tips reflexed. Slightly downy.
BASIN Medium depth. Medium to fairly wide. Regular or slightly ribbed. Ochre or grey russet runs concentrically around basin.
TUBE Funnel-shaped.
STAMENS Median.
CORE LINE Median towards basal.
CORE Median. Axile.
CELLS Obovate or roundish.
SEEDS Large. Obtuse. Broad, Not plump.
LEAVES Medium to small. Broadly oval. Bluntly and broadly serrate. Fairly thin. Slightly upward-folding. Flat. Mid green. Undersides downy.
POLLINATION GROUP 3

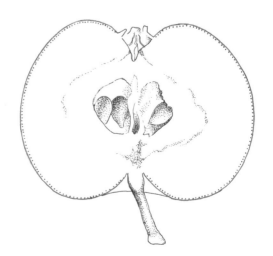

GOLDEN NOBLE

Patrick Flanagan, gardener to Sir Thomas Hare of Stowe Hall, Downham in Norfolk (UK) discovered this apple growing in an old orchard and introduced it to the Horticultural Society of London in 1820. It is a handsome exhibition and garden fruit and one of the best cookers, being acid with an extremely good fruity flavour. The trees are moderately vigorous, upright-spreading and partial tip-bearers. The cropping is moderate. The season is October to December, picking time early October. The flesh is creamy-white, slightly soft, fine-textured and juicy, with very slight aroma.

SIZE Large 80 x 65mm (3^1/8 x 2^1/2"). Medium size on older trees.

SHAPE Round, sometimes flattened and slightly conical. Symmetrical though occasionally lop-sided. Very regular.

SKIN Light green to ochre-green nearest the sun becoming golden yellow. Usually no flush but occasional mottled or striped pinky-brown nearest the sun. Lenticels inconspicuous tiny pink-brown russet dots surrounded by green circle or just pale green dots. Some scarf skin at base noticeable on edges of russet extending from cavity. Skin smooth becoming greasy.

STALK Stout (4mm). Short (9mm). Can be fleshy. Set within cavity.

CAVITY Narrow. Medium depth. Can have fleshy lip on one side.

EYE Medium to small. Slightly to half open. Sepals erect and convergent, broad based with tips reflexed. Fairly downy.

BASIN Rather shallow. Medium width. Slightly ribbed and puckered.

TUBE Long funnel shaped.

STAMENS Marginal.

CORE LINE Faint, median.

CORE Median. Axile closed or open.

CELLS Roundish obovate.

SEEDS Acuminate or acute. Straight. Mid brown. Quite large.

LEAVES Medium size. Acute. Bluntly pointed serrate. Medium thick. Slightly upward-folding, not undulating. Mid green. Undersides downy.

POLLINATION GROUP 4

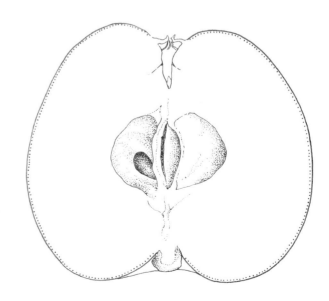

DON'S DELIGHT

A mid to late season culinary apple which was raised in Devon (UK). In the late 70s to early 80s, Alfred Green discovered a tree, thought to be a chance seedling, growing in a neighbour's garden in Torquay and showed it to Don Cockman, a customer advisor to Suttons. Don made grafts and in the 90s brought it to the attention of Kevin Croucher of Thornhayes Nursery. It was introduced in 1996. The trees are large and vigorous, healthy and resistant to scab and canker. They are ideal for growing in the wet west and the cold north and produce very clean fruit in an organic/pesticide free regime. The season is September to January/February, picking time late September/October. The flesh is greenish white, fairly fine-textured, fairly juicy, crisp but not hard and has a sharp rich flavour. Aroma, very slightly acid.

SIZE Medium to large 70–80mm x 70–88mm (3^1/8 x 2^3/4–3^3/8").
SHAPE Round-conical. One or two large well-rounded ribs making one or two pronounced crowns. Irregular. Rounded at base. Usually lop-sided.
SKIN Clear pale green to pale yellowish-green. Can be whitish due to scarf skin. Also some delicate whitish bloom which rubs off. Quarter to half flushed with dull orange-yellow, considerably dotted with tiny dots of scarlet and conspicuous short broken scarlet stripes. On less coloured fruits just the dots and stripes. Lenticels small greenish-grey dots, more conspicuous on pale green skin. Skin smooth and slightly greasy.
STALK Variable. Medium to long. (18–35mm). Fairly slender (3mm). Protrudes beyond base.
CAVITY Fairly wide. Medium depth. Golden brown scaly russet which can streak out over base.
EYE Large. Open. Sepals separated at base, erect with tips slightly reflexed or broken off.
BASIN Irregular. Often very pinched in with some large ribs. Can be some small amounts of brown russet or russet free.
TUBE Cone-shaped.
STAMENS Median.
CORE LINE Median.
CORE Median.
CELLS Obovate. Tufted.
SEEDS Acuminate. Dark brown.
LEAVES Broadly oval. Fairly light brightish green. Thin. Sharply serrate. Undersides very downy.
POLLINATION GROUP 3

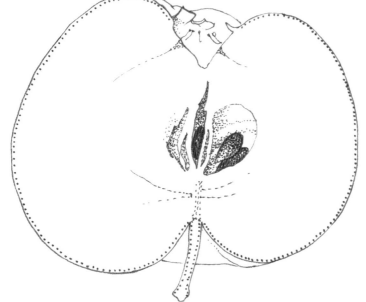

RED FALSTAFF

Red Falstaff is a dessert apple, a more colourful sport of Falstaff (raised in 1965 from James Grieve x Golden Delicious.) It was developed in Kent in the 1980s. The trees are disease and frost resistant and produce vast amounts of blossom making them good pollinators. They are very heavy yielding and the fruits have excellent storing abilities so can be eaten throughout the winter. The season is October to December and beyond. The picking time is mid October.The flesh is cream, fairly coarse-textured, crisp but not hard and juicy. The have a very good flavour, sweet but with a nice acid balance. The skin is a little tough. Fruits have no aroma.

SIZE Medium 72 x 70mm. (2⁷/₈ x 2³/₄").

SHAPE Oblong-conical. Almost no trace of ribs. Regular. Mostly symmetrical.

SKIN Fairly pale greenish-yellow. Half to almost completely covered with pale red to bright red flush on sunny side. Paler flush rather broken and striped, deeper flush more intense with rather indistinct crimson stripes. Can be some broken scarf skin giving the red a milky appearance. Lenticels conspicuous ochre dots on flush and greenish ochre dots on green skin. Some small marks and patches of russet. Skin dry and can be rather hammered. Falstaff is a similar shape and colour but the flush is a rather dull brownish-red with slightly deeper stripes. There is a circle of the deeper red around lenticels giving fruits a rather spotty appearance.

STALK Quite long (28–33mm). Slender (2.5mm). Protrudes well beyond base.

CAVITY Deep and quite narrow. Usually ochre-green and lined with fine slightly scaly pale grey-brown russet which can streak out over base.

EYE Medium size. Partly to fully open. Sepals erect or erect convergent. Tips slightly reflexed or broken off.

BASIN Small to medium.Irregular. Ribbed, sometimes pinched looking, often with fleshy bulge and dimpled. Either greenish-yelow or red with some fine pale ochre russet running concentrically round basin.

TUBE Cone-shaped.

STAMENS Basal.

CORE LINE Basal or towards basal.

CORE Median.

CELLS Round. Axile.

SEEDS Acute. Mid brown. Plump.

LEAVES Medium to small. Acute. Crenate. Mid green. Medium thick. Flat. Slightly upward-folding. Undersides not downy.

POLLINATION GROUP 2. Self fertile.

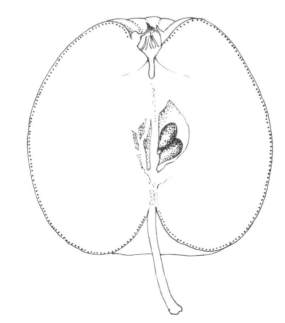

RIBSTON PIPPIN

This highly prized late dessert apple was discovered at Ribston Hall near Knaresborough in Yorkshire and thought to have been raised from seed brought there from Rouen in about 1688, The original tree was blown down in 1810 but, supported by stakes, continued to bear fruit until it died in 1835. From the stem, a young shoot grew into a tree until that too was blown down in 1928. Ribston Pippin received an RHS Award of Merit in 1962. Trees are reasonably hardy, vigorous, upright-spreading and produce spurs very freely. The cropping is moderate to good but fruits are liable to sudden drop at harvest time. Season October to December, picking time early October. The fruits have a rich aromatic flavour with a hint of pear-drops. The flesh is paleyellow, firm, fine-textured and fairly juicy. The skin is a little chewy. Fruits become dry by December and are better eaten earlier.

SIZE Medium large 70 x 58mm (2³/4 x 2¹/4″).

SHAPE Round-conical. Flattened at base and sometimes at apex. Frequently lop-sided. Some unequally large ribs make fruits irregular. Can be flat-sided. Five crowned at apex.

SKIN Greenish-yellow. Quarter to three quarters flushed with orange-brown and numerous broad broken stripes of red. Scratchy patches of green-brown russet at base and on cheeks and some grey-brown russet at apex. Lenticels conspicuous greenish-brown russet dots, pale grey dots at base. Skin smooth and dry becoming greasy.

STALK Medium thick (3mm). Fairly short (10mm). Usually within cavity or level with base.

CAVITY Narrow and deep. Lined with grey-brown scaly russet which can spread over base.

EYE Medium to small. Closed or slightly open. Sepals erect connivent with tips reflexed. Very downy.

BASIN Medium to deep and fairly wide. Ribbed and puckered. Occasional trace of beading. Some cinnamon or grey-brown russet.

TUBE Long cone or funnel-shaped.

STAMENS Median or basal.

CORE LINE Median towards basal. Usually joins below stamens.

CORE Median. Axile. Sometimes slightly open.

CELLS Obovate.

SEEDS Few. Long oval. Acuminate. Fairly plump.

LEAVES Medium size. Broadly oval. Serrate, Medium thick. Flat. Upward-folding. Mid green. Undersides downy.

POLLINATION GROUP 2 Triploid.

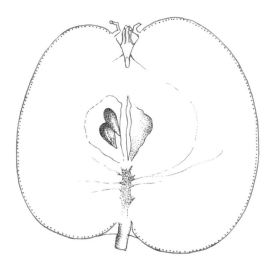

HOARY MORNING

This attractive mid to late dual purpose apple was raised in the UK, thought to be in Somerset, in the early 1800s. It was first recorded in 1819. The trees are moderately vigorous, upright-spreading and produce spurs fairly freely. It is very resistant to scab and crops consistently well. The season is October to January, picking time mid September. The flesh is greenish-yellow, fairly coarse-textured and rather dry. It has moderate flavour when cooked, slightly acid with some sweetness and keeps its shape. When eaten raw, they are rather sour with not much flavour. Fruits have no aroma.

SIZE Medium 74 x 52mm ($2^7/8$ x $2^1/8$").
SHAPE Flat to round-conical. Broad flat base. Mostly regular with faint broad ribs.
SKIN Pale yellow-green becoming yellow. Three quarters to almost completely covered with long broad brownish-red to bright red stripes which are sometimes broken. Initially covered with heavy white bloom which rubs off when handled. Scarf skin at base. Lenticels distinct large pale grey dots at base otherwise tiny grey-brown dots. Skin slightly greasy.
STALK Short to medium. Stout. Level with base or protruding slightly beyond.
CAVITY Wide. Medium depth. Some golden russet.
EYE Medium. Closed or slightly open. Sepals broad, connivent with tips reflexed.
BASIN Narrow. Fairly shallow. Distinctly ribbed and pinched looking. Sometimes beaded.
TUBE Cone-shaped.
STAMENS Marginal.
CORE LINE Median.
CORE Median. Axile.
CELLS Ovate.
SEEDS Quite large and numerous. Acute.
LEAVES Medium to small. Oval to broadly oval. Serrate. Fairly thin. Fairly bright green. Undulating. Undersides slightly downy.
POLLINATION GROUP 4

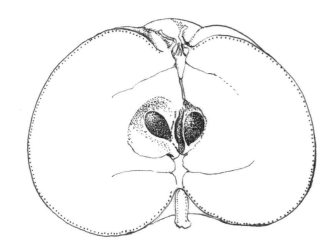

GALA

This mid to late season dessert apple is a fairly new introduction. It was raised in New Zealand in 1934 by J.H.Kidd of Greytown, Wairarapa from Kidd's Orange Red x Golden Delicious. It was selected in 1939, introduced in 1960 and named in 1965. Commercially its popularity is open to speculation because of the size and indefinite colour of fruit; however, it is widely available. The flavour is erratic: it can be very sweet and interesting with a hint of pear-drops but can be rather flat in some seasons and the flavour fades with keeping. The season is October to early January, picking time early October. The cropping is good and heavy.The flesh is yellowish-cream, firm but quite soft, fine-textured and The trees are moderately vigorous, spreading and produce spurs freely. They are suitable for the north but prone to scab. Royal Gala is a highly coloured cultivar from a sport of Gala, introduced in the 1970s.

SIZE Medium 61 x 58mm ($2^3/8$ x $2^1/4$").
SHAPE Oblong-conical. Distinct narrow ribs noticeable at apex where they terminate in five rather flat crowns. Symmetrical or lop-sided. Regular.
SKIN Pale green becoming bright yellow. Quarter to almost completely flushed with dense crimson dots and flecks under which the skin is yellow or orange-red. Short narrow broken stripes of crimson. Lenticels conspicuous green or grey-brown russet dots. Some small patches of ochre russet.Thin purplish scarf skin at base. Skin smooth and dry.
STALK Slender. (2–2.5mm). Very long (25mm). Set at an angle.
CAVITY Wide and fairly deep, sometimes lipped. Partly lined with green-gold russet overlaid with fine grey scaling.
EYE Medium size. Closed or slightly open. Sepals very long, broad based, tapering to a sharp point, convergent with most tips reflexed. Sepals green with brown tips. Slightly downy.
BASIN Wide and deep. Pronounced ribs. Slightly puckered. Occasional small patches of grey russet.
TUBE funnel-shaped.
STAMENS Median.
CORE LINE Indefinite. Basal.
CORE Median. Axile, sometimes slightly open.
CELLS Ovate
SEEDS Obtuse. Quite large, fairly plump. Regular.
LEAVES Medium to small. Narrow acute. Serrate. Rather thin. Flat not undulating. Upward-folding. Dark green. Not downy.
POLLINATION GROUP 4

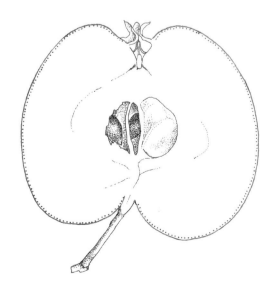

SWEET SOCIETY

This is a new dessert apple raised by Hugh Ermen in 2004 and was part of a range of plants selected by the RHS to celebrate their Bicentennial in that year. The trees are moderately vigorous, upright spreading and become compact bushy trees which spur freely. The cropping is heavy and thinning is recommended to avoid small fruits. The season is October to January, picking time late September. The flavour is highly aromatic, sweet, but with a good fresh sharpness. The flesh is creamy-white, fine-textured, firm but not hard and fairly juicy. A refreshing and tasty apple.

SIZE Medium 68 x 60mm (2⁵/8 x 2¹/4").
SHAPE Round-conical. Regular. Slight trace of one well rounded rib. Mostly symmetrical.
SKIN Bright yellow. Almost to completely covered with brilliant red and crimson flush with speckled red over remaining yellow skin. Many short broken stripes and dots of deep crimson. Lenticels very conspicuous small yellow or pale green dots. Can be some grey scarf skin at base. Some tiny flecks of grey-brown or ochre russet. Skin smooth and dry.
STALK Stout (4mm). Long (30mm). Protrudes well beyond base.
CAVITY Medium to shallow depth. Medium width. Green with some fine grey russet radiating out sometimes over base.
EYE Medium size. Sepals short, erect convergent. Tips slightly reflexed or broken off.
BASIN Medium width. Fairly shallow. Trace of ribs and almost beaded. Downy. Can be some patchy golden russet.
TUBE Cone-shaped.
STAMENS Median.
CORE LINE Basal.
CORE Distant. Axile.
CELLS Round.
SEEDS Medium size. Mid brown and fairly plump.
LEAVES Small. Fairly dark green. Acute. Crenate. Thick and leathery. Fairly flat. Slightly upward-folding. Undersides slightly downy.
POLLINATION GROUP 4

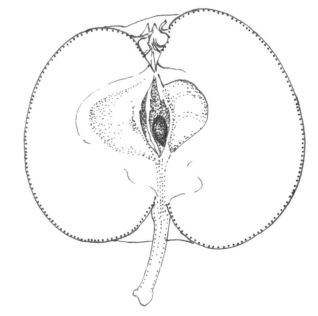

GASCOYNE'S SCARLET

A very attractive mid to late season dual purpose apple which was raised by Mr. Gascoyne of Bapchild Court, Sittingbourne, Kent (UK). It was introduced in 1871 by G. Bunyard & Co of Maidstone and received an RHS First Class Certificate in 1887. The trees are very vigorous, rather gaunt, upright-spreading and partial tip-bearers. The cropping is moderate and the season is October to January, picking time mid September. The fruits are sweet with a refreshing acidity with greenish white flesh, cream near the flush, firm, fine-textured and juicy. When cooked the flesh is pale greenish-yellow, sub acid with quite a good flavour. It breaks up completely though not to a fluff.

SIZE Large 77 x 64mm (3 x 2½").

SHAPE Flat-round. Flattened at base and apex. Five crowned at apex. Large prominent well rounded ribs. Symmetrical. Regular.

SKIN. Pale yellow-green to pale primrose yellow. Half to almost completely covered with bright crimson flush. Some indistinct short broken stripes of deep crimson. Lenticels very pronounced green or ochre russet dots some being quite large. The quantity of green lenticels make the yellow skin appear green. Skin smooth and slightly greasy.

STALK Medium to fairly stout (3–3.5mm). Medium to long (15–25mm). Stalk set so deep into cavity it is level with base or protruding slightly beyond.

CAVITY Deep and fairly narrow. Regular cone-shape. Dark ochre green with variable scaly grown russet which can streak over base.

EYE Large. Partly open. Sepals broad-based, long and sharply pointed. Erect convergent with tips reflexed.

BASIN Wide and very deep. Prominently ribbed and puckered. Downy

TUBE Cone-shaped.

STAMENS Median towards basal.

CORE LINE Basal clasping, sometimes joining.

CORE Median to somewhat distant. Axile or abaxile.

CELLS Obovate.

SEEDS Obtuse. Oval and fairly plump. Regular.

LEAVES Large. Broadly oval. Serrate. Medium to thin. Slightly undulating, sometimes downward-folding. Dark green. Undersides downy.

POLLINATION GROUP 5. Triploid.

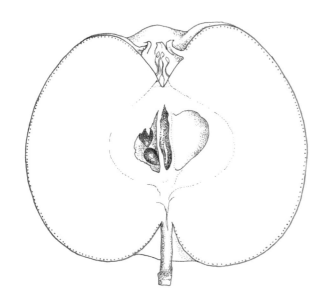

HEREFORDSHIRE RUSSET

This is a new dessert variety bred by Hugh Erwin at the turn of this century. The parentage is unknown. It is a small but delicious golden russet with the aromatic flavour of the Cox. The trees are strong, upright-spreading and suitable for growing in the northern and wetter regions of the country: they are spur-bearers and heavy croppers.The season is October to January, picking time late September. The flesh is creamy-white tinged slightly green, fine-textured, firm, quite hard and slightly juicy when cut. I tasted them at the end of September and they were crisp and juicy with a lovely fruity flavour, slightly acid but with plenty of sweetness. The skin was not tough. No aroma.

SIZE Medium 70 x 65mm (2³/4 x 2⁵/8").
SHAPE Conical. Mostly symmetrical. Regular. Some slight crowns at apex.
SKIN Fairly bright yellowish green. Partly to three quarters covered with fine golden sometimes netted russet which is overlaid in parts with some pale grey-brown slightly scaly russet. Lenticels very conspicuous raised pale grey-brown russet dots. Skin slightly rough and dry.
STALK Long 23mm. Medium thick (3.5mm). Protrudes well beyond base.
CAVITY Fairly deep and narrow. Can be bright green but usually russetted.
EYE Medium size. Closed. Sepals long, connivent with tips well reflexed or broken off.
BASIN. Medium depth. Quite narrow. Trace of ribs. Some fruits have one or more dimples around the crowns. Usually bright green around the eye.
TUBE Cone-shaped.
STAMENS Basal.
CORE LINE Basal sometimes clasping.
CORE Medium towards sessile.
CELLS Round or obovate.
SEEDS Acute. Dark brown. Plump. Numerous.
LEAVES Acute. Mid green. Medium size. Slightly bluntly serrate. Upward-folding. Fairly flat. Undersides quite downy.
POLLINATION GROUP 3. Partly self-fertile.

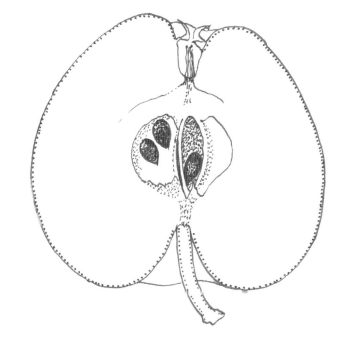

FIESTA

This attractive, brightly coloured apple is a cross between Cox's Orange Pippin and Idared, a new introduction from East Malling Research Station in Kent (UK) and previously known as T31/31.

The flavour is much like the Cox and is proving to be of great commercial interest because it crops consistently and stores well. It received a Preliminary Commendation from the RHS in 1987. Fiesta shows a degree of self-fertility and the flowers withstand low temperatures well. Trees are moderately vigorous, upright-spreading and spur-bearers. The season is late October to January, picking time mid September. The flesh is yellowish, rather coarse-textured, firm, crisp and juicy. The flavour is rich and vinous, quite sweet with a nice tang making it refreshing. Fruits have a very slight rather acid aroma.

SIZE Medium small 58 x 51mm (2¼ x 2") to medium 67 x 51mm (2⅝ x 2").

SHAPE Round to flat-round. Usually slightly lop-sided. No ribs or a slight trace of one or two well rounded ones. Regular.

SKIN Greenish-yellow becoming yellow. Up to three quarters covered with bright red flush which can be a dense crimson on sunny side or a thinner speckled red on shaded side. Short broken stripes of crimson to purplish-crimson on brighter flush. Lenticels conspicuous whitish dots. Occasional small ochre or grey-bown russet patches. Skin smooth becoming greasy.

STALK Fairly slender to medium stout (2.5–3.5mm). Quite long (22–30mm). Frequently set at an angle.

CAVITY Fairly wide. Medium depth. Frequently lipped. There can be some scaly brown russet.

EYE Medium size. Closed or slightly open. Sepals short, erect or flattish convergent with tips reflexed or broken off.

BASIN Medium width and fairly shallow. Slightly puckered. Very downy. Occasional small amounts of brown russet.

TUBE Funnel-shaped.

STAMENS Marginal.

CORE LINE Faint. Median towards marginal.

CORE Median. Axile.

CELLS Round or roundish obovate. Slightly tufted.

SEEDS Large. Obtuse. Flat on one side. Curved.

LEAVES Medium to small. Acute to narrow acute. Serrate. Medium thick. Flat or undulating. Upward-folding. Dark blue-green. Undersides downy.

POLLINATION GROUP 3

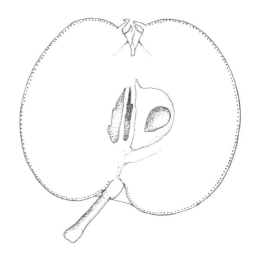

SCOTCH BRIDGET

This old culinary variety, also known as White Calville, originated in Scotland in 1851. It is commonly found in Cumbria and seems to thrive in northern locations where it will produce regular heavy crops even after a harsh winter, late frosts and a poor summer. The parents are unknown. The season is October to December at which time the fruit ripens to become a dessert apple. The picking time is early October. The trees have moderate vigour, are upright-spreading and spur-bearers. The flesh is cream tinged green, fine-textured, firm and rather dry. When cooked it does not break up but stays whole, cooks to a cream colour and doesn't need much sugar. It has a good fairly delicate flavour. The skin is a little tough. Fruits have a slight aroma.

SIZE Medium large 75–80 x 65–70mm (3–3^{1}/$_{4}$ x 2^{1}/$_{2}$–2^{3}/$_{4}$").
SHAPE Round-conical. Large unequal ribs making fruits often odd shaped and irregular. Frequently lop-sided. Five-crowned at apex.
SKIN Bright mid green becoming dull golden yellow. Slightly to three quarters covered with dull orange to dull red flush overlaid with dots and mostly broken stripes of red and crimson. A few small scuffs and patches of greenish-grey russet. Some fruits have defined patches of fine golden russet like a separate layer stuck on the fruit. Lenticels inconspicuous greenish-grey russet dots. Skin smooth and slightly greasy. Can have russetted hair line.
STALK Short (15mm) and fairly stout (4–5mm). Often fleshy. Within cavity.
CAVITY Medium width. Medium depth. Sometimes fleshy bulge on one side making it narrow. Usually some grey-brown russet present and occasionally some scaly brown russet.
EYE Medium size. Slightly open. Sepals connivent with tips reflexed.
BASIN Fairly shallow. Medium width to narrow. Ribs visible and often beaded. Russet free. Very slightly downy.
TUBE Cone-shaped.
STAMENS Median.
CORE LINE Basal or towards basal.
CORE Median.
CELLS Obovate. Tufted.
SEEDS Acuminate. Dark brown. Not very plump.
LEAVES Oval. Fairly light grey-green. Serrate. Upward-folding. Slightly undulating. Undersides very downy.
POLLINATION GROUP 3

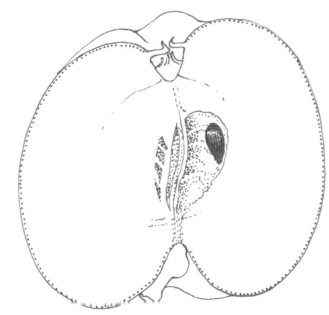

CATSHEAD

This is an old English culinary apple which was known in the 1600s, but possibly originated earlier. An apple with this name is in the national fruit trials collection but it is not certain whether it is the true cultivar. It was widely grown in the nineteenth century and popular for apple dumplings. The trees are of medium vigour, very spreading and produce spurs fairly freely. The season is late October to January, picking time ate September to early October. The fruit cooks to a fluff with a good flavour, sweet but with a good balance of acid. The flesh is creamy-white tinged green, fairly fine-textured, firm and fairly juicy.

SIZE Very large 90 x 85mm (3³/4 x 3³/8").

SHAPE Oblong to oblong-conical. Irregular. Often lop-sided. Large well rounded ribs often making it flat-sided. Distinctly five crowned at apex. Invariably angular and odd shaped.

SKIN Fairly dull yellowish-green. Considerably covered with silvery scarf-skin with broken stripes missing showing green skin beneath giving the fruit a slightly striped appearance. Can be a russet hair line. A few grey-brown lenticel dots. Some fruits slightly flushed dull orange. Skin smooth becoming greasy.

STALK Short (10mm). Stout (3mm). Within cavity.

CAVITY Fairly deep. Medium to narrow. Somewhat irregular. Some greenish-brown russet within.

EYE Large. Open. Sepals short, separated, broad-based tapering to reflexed pointed tips.

BASIN Irregular. Medium width. Fairly deep. Distinctly ribbed and puckered. Some small patches of brown russet within.

TUBE Cone-shaped.

STAMENS Median towards marginal.

CORE LINE Basal meeting or towards median.

CORE Median to sessile. Axile open.

CELLS Round. Tufted.

SEEDS Acute

LEAVES. Small to medium. Olive green. Broadly oval. Slightly undulating. Upward-folding. Medium thick. Serrate to bluntly serrate. Undersides very downy.

POLLINATION GROUP 3

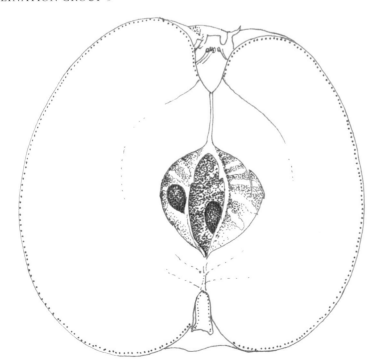

MARGIL

The history of this old dessert apple is uncertain. One theory is that it was brought to England by George London who worked in the gardens at Versailles under De La Quintinye and who was a partner of the Brompton Park nursery where this apple was extensively cultivated as early as 1750. Rogers in his *Fruit Cultivator* of 1834 claims its reputation as a dessert fruit for seventy years, and the first time he saw it was an espalier planted by Sir William Temple in the Sheen Garden. They make small trees of weak growth well suited to small gardens. They are hardy but susceptible to late frost due to the early flowering. They are spur-bearers, the cropping is good and the season is October to January, picking time early October. The fruits are rich and sweet, the flesh creamy white, firm, slightly coarse-textured but rather dry. They are sweetly aromatic.

SIZE Small 54 x 51mm (2¹/8 x 2″).
SHAPE Round-conical. Symmetrical or lop-sided. Prominent irregular ribs making it angular and flat-sided. Irregular. Five-crowned.
SKIN Greenish-yellow to yellow. Half to three quarters flushed orange-red, to crimson on well coloured fruits. Distinct broad, broken or long stripes of crimson, paler over the yellow skin. Some fine ochre or green-grey russet patches. Lenticels conspicuous as grey-brown or ochre russet dots. Skin smooth and slightly greasy.
STALK Slender (2mm). Medium length (13–18mm). Protrudes beyond base.
CAVITY Narrow and deep with stalk deeply inserted. Sometimes lipped. Lined with grey-green scaly russet which radiates out.
EYE Fairly small. closed or partly open. Sepals broad-based, erect and rather pinched together.
BASIN Narrow. Medium depth. Distinctly ribbed, sometimes beaded. Often falls away on one side. Usually some fine grey-brown russet.
TUBE Deep. Cone or slightly funnel-shaped.
STAMENS Median.
CORE LINE Median sometimes inclined towards basal.
CORE Median. Axile.
CELLS Roundish ovate or round.
SEEDS Acuminate or acute. Quite large. Oval. Pointed or blunt.
LEAVES Medium to small. Acute to narrow acute. Crenate. Medium thick. Upward-folding. Flat. Mid grey-green. Moderately downy,
POLLINATION GROUP 2

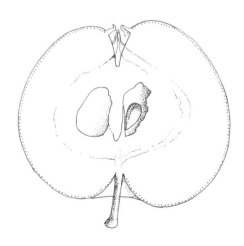

SPARTAN

This McIntosh type dessert apple is of Canadian origin. It was raised in 1926 at the Dominion Experiment Station, Summerland, British Columbia, by R.C.Palmer from McIntosh x Yellow Newtown Pippin. The original tree first fruited in 1932 and it was introduced in 1936. It is fairly important commercially. The trees are of moderate vigour, upright-spreading and spur-bearers, producing spurs freely. They are rather prone to canker but will tolerate growing in the North and West. The season is October to February, picking time early October. The fruits are small unless thinned and the trees fed generously. The flavour is rich, sweet and juicy but I found them rather dry by January. The flesh is white, firm, fine-textured. The skin is a little tough. They have a strong sweet aroma.

SIZE Medium 64 x 58mm (2¹/₂ x 2¹/₄").
SHAPE Round conical. Frequently lop-sided. Indistinct ribs, more distinct towards apex. Often five crowned at apex. Regular.
SKIN Pale green becoming pale yellow. Three quarters to almost completely flushed crimson or dense purple-crimson with indistinct thin broken stripes of purple or brownish-crimson. Considerably bloomed. Some streaky scarf skin at base. Lenticels distinct tiny white or ochre russet dots on flush, tiny green-grey or pink dots on yellow or green skin. Skin very smooth and dry.
STALK Variable. Fairly slender (2.5mm) to medium or stout (3–4mm). Medium length (15–20mm).
CAVITY Deep and rather narrow. Often lipped. Sometimes some fine green-ochre russet.
EYE Medium to small. Closed or partly open. Sepals erect, can be convergent and separated at base, tips reflexed. Very downy.
BASIN Medium width and depth. Can be pinched looking. Ribbed and puckered, sometimes beaded.
TUBE Deep slightly funnel almost cone-shaped.
STAMENS Median.
CORE LINE Almost basal, clasping.
CORE Median. Axile or abaxile.
CELLS Obovate.
SEEDS Acute or acuminate.
LEAVES Medium size. Oval. Serrate. Medium thick. Flat not undulating. Mid grey-green. Undersides very downy.
POLLINATION GROUP 3

CORTLAND

Plant breeders at the New York State Agricultural Experiment Station in Geneva, New York began breeding McIntosh with other varieties and one of the first was a cross with Ben Nevis which produced Cortland. This late dessert apple was developed in 1898 and named after nearby Cortland County, New York. It was introduced in 1900. The trees are of medium size, upright-spreading and produce spurs fairly freely. Season mid October to February. Picking time late September. The flesh is very white, fine textured and juicy. It stays white a long time after cutting. The flavour is scented and reminded me of a sweet pineapple. It is fairly mild but refreshing with slightly tough skin.

SIZE Medium 75 x 60mm (3 x 2³/8").
SHAPE Flat-round to round-conical. Can be a very slight trace of well rounded ribs otherwise regular. Can be slightly lop-sided.
SKIN Fairly bright green becoming pale yellow. Half to completely covered with crimson to deep crimson flush. Slight trace of broken crimson stripes only visible on paler flush. Fruit covered with fine bloom on tree which rubs off and fruit polishes to a fine shine. Lenticels fairly conspicuous small pale russet dots. Skin smooth and dry.
STALK Medium width to stout (3–5mm). Fairly short (17mm). Within cavity or protruding slightly beyond. Can have fleshy knob at base setting stalk at an angle.
CAVITY Fairly deep and wide. Usually lined with ochre-green which can be overlaid with golden scaly russet. Can extend over base.
EYE Small to medium size. Tightly closed or partly open. Sepals small, erect, tips slightly reflexed.
BASIN Medium depth. Wide. Ribbed. Often a small dimple. Can be irregular. No russet.
TUBE Cone shaped.
STAMENS Median.
CORE LINE Median to basal.
CORE Median to distant. Abaxile.
CELLS Obovate.
SEEDS Numerous. Dark brown. Small. Fairly plump. Obtuse.
LEAVES Medium size. Crenate. Acute. Undulating. Slightly upward-folding. Fairly thick and leathery. Dark olive green. Undersides downy.
POLLINATION GROUP 3

LAXTON'S SUPERB

This late dessert apple was raised in England in 1897 by Laxton Bros Ltd of Bedford (UK) from Wyken Pippin x Cox's Orange Pippin. It received an RHS Award of Merit in 1919 and a First Class Certificate in 1921. It was introduced in 1922 and is grown commercially today. Trees are vigorous, upright-spreading and spur-bearers. The young trees crop regularly and heavily but on becoming established tend to become biennial and the fruit can vary considerably in flavour from season to season and area to area. They can have an almost Cox-like flavour in hot summers.The trees are hardy and suitable for the north but are prone to scab. The season is November to January, picking time early October. The flesh is white tinged slightly green, fine-textured, firm but tender and juicy. Fruits have no aroma and skin is tough.

SIZE Medium to medium large 67 x 61mm ($2^5/8$ x $2^3/8$") or 70 x 61 ($2^3/4$ x $2^3/8$").

SHAPE Round-conical to conical. Indistinct well rounded ribs. Symmetrical or slightly lop-sided. Fairly regular.

SKIN Pale greenish-yellow. Half to almost completely covered with dull purplish-red, less dense and mottled away from the sun. A few short broken stripes of dark purple-crimson. Some small dirty green-ochre russet patches. Lenticels conspicuous quite large grey-brown dots, sometimes star-shaped. There can be some fine, slightly broken scarf skin. Skin dry and slightly rough. Often confused with Allington Pippin.

STALK Medium thick (3mm) and medium to fairly long (15–20mm). Protrudes well beyond base.

CAVITY Medium width and depth. Regular. Lined with ochre-green over which there is some scaly dark grey-brown russet which can come out over the shoulder. Sometimes lipped on one side.

EYE Quite large. Open or partly open. Sepals fairly broad based, erect with tips reflexed or broken off.

BASIN Medium width and depth. Regular and slightly puckered.

TUBE Cone-shaped.

STAMENS Median.

CORE LINE Basal, clasping.

CORE Median. Axile.

CELLS Ovate, or roundish.

SEEDS Acuminate. Fairly plump. Regular.

LEAVES Medium size. Acute. Sharply, bluntly or broadly serrate. Fairly thin. Not undulating. Upward-folding. Mid grey-green. Downy.

POLLINATION GROUP 3

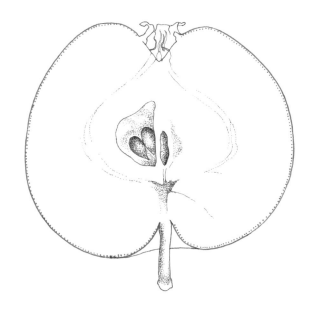

COX'S ORANGE PIPPIN

The man who raised this historic fruit was Richard Cox (1777-1845), a retired brewer from Bermondsey, London. It was raised from pips of a Ribston Pippin at Colnbrook Lawn, near Slough, Bucks (UK) in about 1825 where the original tree grew until destroyed by a storm in 1911. The fruit was introduced by Charles Turner in about 1850 and it received an RHS Award of Merit and a First Class Certificate in 1962. Trees are moderately vigorous, upright-spreading and spur-bearers. They are prone to canker in cold wet areas and susceptible to scab and mildew. They prefer a warm climate and need good soil conditions and favourable environment to crop well. The season is late October to January, picking time early to mid October. The fruit have a very rich flavour, sweet, slightly sharp, nutty and aromatic. The flesh is creamy-white, firm, fine-textured and juicy.

SIZE Medium 64 x 54 (2^{1}/$_{2}$ x 2^{1}/$_{8}$").
SHAPE Round-conical. Slight trace of well-rounded ribs. Symmetrical. Slightly flattened at base and apex. Regular.
SKIN Yellowish-green becoming clear yellow. Up to three quarters flushed with orange-red to deeper red. Some broken stripes of brownish-crimson. Some patches and dots of fine grey-brown russet. Some patchy scarf skin towards base. The fruits brighten in colour as they mature. Skin dry and fairly smooth.
STALK Fairly slender to medium (2.5–3mm). Medium length (15–20mm). Extends beyond base.
CAVITY Fairly wide, medium depth. Often lipped. Lined with ochre-green and scaly grey-brown russet usually extending over base.
EYE Fairly small. Half open. Sepals rather narrow, erect convergent, three quarters reflexed. Stamens often present. Fairly downy.
BASIN Wide and shallow. Slightly ribbed. Some grey-brown russet.
TUBE Funnel-shaped
STAMENS Median.
CORE LINE Basal, clasping.
CORE Median. Axile.
CELLS Obovate
SEEDS Obtuse. Wide. Large for size of apple.
LEAVES Medium size. Acute. Bluntly serrate. Medium to thin. Slightly upward-folding. Mid yellow-green. Undersides downy.
POLLINATION GROUP 3

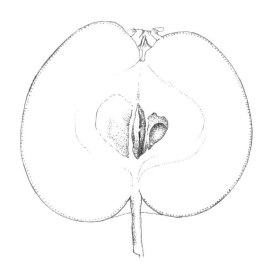

HOWGATE WONDER

A very large late culinary apple which was raised in 1915-16 by Mr. G Wratton of Howgate Lane, Bembridge, Isle of Wight. It was introduced in 1932 by Stuart Low & Co. of Enfield in Middlesex and received an RHS Award of Merit in 1949. It is a heavy cropper when fully established producing an attractive fruit that stores well. It is a useful pollinator for Bramley's Seedling. The trees are vigorous, produce spurs feely and, being hardy, are suitable for growing in the north of the UK. The season is October to March, picking time early October. The fruit is only fair in quality, sub-acid with not much flavour. The flesh is creamy-white fine-textured and firm and becomes a rather drab yellow when cooked and breaks up almost completely. Almost no aroma.

SIZE Very large 86 x 72mm (3³/8 x 2⁷/8").

SHAPE Round-conical. Flattened at base tapering to a flattened apex. Distinct broad ribs, sometimes angular which terminate in five pronounced crowns at apex. Can be flat-sided. Regular. Symmetrical or lop-sided.

SKIN Light yellow-green to greenish-yellow. Up to half flushed with thin orange-brown with short broken stripes of the same colour or a darker greyed-red. Usually some patchy scarf skin at base or on cheeks making the flush appear more purple-grey. Usually no russet. Lenticels indistinct white or grey dots with circle of scarf skin, more numerous at apex and entering basin. Skin very smooth and dry.

STALK Short (8–13mm). Stout (5mm). Often fleshy. Within cavity

CAVITY Wide and deep. Frequently lipped with stalk enveloped in the lip and sunk deep into cavity. Usually green with conspicuous dots of scarf skin. Partly or completely lined with light brown russet which can streak over base.

EYE Large. Closed or half open. Sepals convergent, broad at base with tips reflexed. Very downy.

BASIN Wide and deep. Ribbed and puckered. Irregular.

TUBE Funnel-shaped.

STAMENS Median.

CORE LINE Median. Sometimes two.

CORE Median. Axile.

CELLS Obovate. Sightly tufted.

SEEDS Acute. Broad and fairly plump. Straight.

LEAVES. Medium. Broadly acute. Serrate. Medium thick. Slightly undulating. Dark green. Undersides downy.

POLLINATION GROUP 4

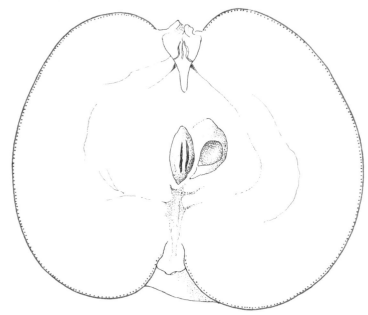

WINTER GEM

This is a fairly new introduction raised in 1975 by Hugh Ermen in Faversham in Kent (UK) from Cox's Orange Pippin x Grimes Golden. It was introduced by F.P.Matthews of Worcestershire in 1993. It is a good dessert garden variety. The trees are upright-spreading, moderately vigorous, spur-bearers, and crop heavily. The season is October to March, picking time early October. The fruits are firm, crisp and juicy and have a very rich aromatic flavour. The flesh is creamy-yellow and is fairly coarse-textured. The fruits have a slight aroma.

SIZE large 75–80 x 70mm (3–3^{1}/4 x 2^{3}/4").
SHAPE Oblong to oblong-conical. Large well-rounded ribs which can make fruits look squarish. Often slightly lop-sided. Irregular.
SKIN Golden yellow. Half to three quarters covered with mottled orange-red flush which is overlaid with lots of short broken stripes and speckles of scarlet. Some scattered patches and netting of golden-brown russet which spreads from base, giving the skin a slightly rough feel, otherwise smooth and dry. Lenticels prominent at base and over yellow skin as grey-brown dots, tiny grey-brown dots over dense flush. A glowing and attractive fruit.
STALK Stout (4mm) and shortish (14mm). Within cavity or protruding slightly beyond. Usually set at an angle and often with fleshy knob.
CAVITY Shallow to medium depth. Fairly narrow. Covered with greenish-ochre russet, frequently overlaid with dots and tiny patches of brown russet which spreads out over base.
EYE Large. Open. Sepals broad based, erect. Tips usually reflexed.
BASIN Narrow, rather irregular with ribs visible. Frequently a little dimple on one side. Varying amounts of grey-brown russet within.
TUBE Funnel-shaped.
STAMENS Median.
CORE LINE Basal, clasping.
CORE Median. Axile.
CELLS Obovate.
SEEDS Acute.
LEAVES Medium size. Acute. Serrate. Fairly flat. Medium thick. Undersides fairly downy.
POLLINATION GROUP 3

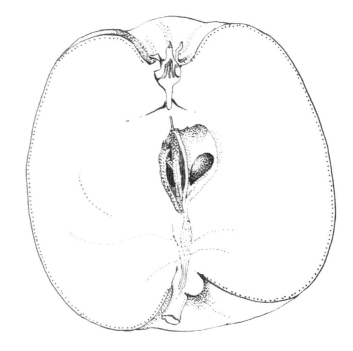

UPTON PYNE

This apple was raised in Devon (UK) by George Pyne of Topsham and was introduced in 1910 and exhibited at the RHS in 1933. The parentage is unknown. It is a large mid season dual purpose apple, primarily an exhibition and dessert variety. The trees are upright and of medium vigour. The season is November–February, picking time late September to early October. The flesh is creamy-white, firm, coarse-textured, juicy and somewhat acid. It cooks to a puree with a distinct slightly pineapple flavour. As a dessert it has moderate flavour. Aroma almost none, slightly acid when cut.

SIZE Large 85 x 77mm (3¼ x 3″).
SHAPE Conical. Very broad at base. Well rounded ribs making it occasionally flat-sided. Slightly five-crowned at apex.
SKIN Pale greenish-yellow becoming yellow to golden yellow facing the sun. Quarter to half covered with broad broken stripes and dots of pale red. Scarf skin, mostly at base. Lenticels inconspicuous tiny green-ochre dots. Skin smooth and dry
STALK Short(10mm). Slender to medium (3mm). Usually within cavity.
CAVITY Wide. Medium depth. Small amounts of grey russet.
EYE Medium. Open. Sepals broad based, short, often separated at base, tips reflexed.
BASIN Medium width and depth. Ribbed. Sometimes beaded. Little or no russet.
TUBE Funnel-shaped.
STAMENS Median.
CORE LINE Median.
CORE Median.
CELLS Obovate, tufted. Axile.
SEEDS Acute.
LEAVES Broadly acute. Yellow-green. Medium thick. Serrate to bi-serrate. Undersides fairly downy.
POLLINATION GROUP 3. Self sterile.

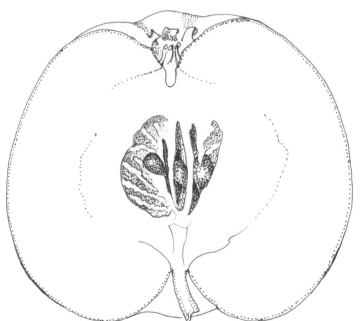

SUNTAN

A good new late dessert apple raised in 1955 by Dr. Alston of East Malling Research Station, Maidstone,Kent (UK) from Cox's Orange Pippin x Court Pendu Plat. Because of its late flowering, later than Cox, it is useful where frosts are a problem. It is triploid but, in difficult areas with suitable pollinators and good management, it could produce a crop where other varieties may fail. The trees are very vigorous and a graft on to a dwarfing rootstock is recommended for the small garden or restricted form. Trees are wide-spreading and produce spurs freely. The season is November to January and the cropping is good. Picking time mid October. The flavour is richly sweet and very acid, mellowing by December. The flesh is yellowish-cream, firm and fairly juicy, rather coarse-textured yet crisp. No aroma.

SIZE Medium large 70 x 54mm (2³/₄ x 2¹/₈″).
SHAPE Flat-round. Distinctly flattened at base and apex. Slight trace of ribs. Symmetrical or slightly lop-sided. Fairly regular.
SKIN Rather dull greenish-yellow. Quarter to three quarters flushed dull orange-red. Indistinct short, fairly broad, broken stripes of dull crimson. Lenticels very conspicuous large ochre or grey-brown russet dots with similar coloured patches of russet making surface textured, otherwise smooth. Dry becoming slightly greasy.
STALK Fairly stout to stout (3.5–4.5mm) Medium length (15–20mm) protruding beyond base, or short (8mm) and level with base.
CAVITY Medium width and depth. Regular. Cavity greenish-ochre overlaid with grey-brown slightly scaly russet which streaks out over base.
EYE Large, half open. Sepals broad based, erect or flattish convergent with tips reflexed or broken off. Fairly downy.
BASIN Wide and shallow. Slightly puckered with some beading. Sometimes lined with fine grey-brown russet which can extend over apex.
TUBE Short, broad funnel-shaped.
STAMENS Median.
CORE LINE Rather faint. Median joining at base of funnel bowl.
CORE Somewhat sessile. Axile.
CELLS Obovate.
SEEDS Large. Obtuse. Quite plump. Regular.
LEAVES Quite large. Rather long. Acute. Bluntly serrate. Medium thick. Flat not undulating. Upward-folding. Mid grey-green. Very downy.
POLLINATION GROUP 5. Triploid.

KIDD'S ORANGE RED

A mid to very late dessert apple raised in New Zealand in 1924 by J.H.Kidd of Greytown, Wairarapa from Cox's Orange Pippin x Delicious. It is grown commercially in New Zealand. It was introduced into England about 1932 and gained popularity as a supplement to Cox's Orange Pippin due to its prolonged storage time but it is prone to excessive russetting in some districts and is susceptible to damage from certain sprays. There is a highly coloured, good flavoured sport available called Captain Kidd. The trees are moderately vigorous, upright-spreading and produce spurs quite freely. The season is November to January. The cropping is good and picking time is early October. It is one of the finest flavoured dessert apples, sweet and aromatic but fruits can be small unless thinned. The flesh is creamy-white, fine-textured, firm and juicy. Fruits have a slight vinous aroma.

SIZE Medium 67 x 64 (2⅝ x 2½").

SHAPE Conical. Rounded at shoulder but flattened at base. Narrow at apex. Fairly distinct ribs and sometimes flat-sided towards apex. Slightly five-crowned at apex. Fairly regular.

SKIN Pale greenish-yellow becoming pale yellow. Half to almost completely covered with crimson flush which is more red than crimson at the edges. Indistinct narrow broken stripes of purplish-crimson. Some small patches of pale grey-brown or greenish-ochre russet. Some indistinct scarf skin at base. Lenticels inconspicuous grey-brown russet dots. Skin smooth and dry.

STALK Fairly stout (3.5–4mm). Medium to fairly long (15–20mm). Protrudes just or well beyond base. Often fleshy at spur attachment.

CAVITY Wide and deep. Partly or completely lined with golden-grey russet which can streak over base and scatter over cheeks.

EYE Small to medium. Closed. Sepals erect convervent with tips reflexed. Very downy.

BASIN Medium width and shallow. Ribbed. Some fine brown russet present. Sometimes downy.

TUBE Long cone-shaped or funnel-shaped.

STAMENS Median.

CORE LINE Rather faint. Basal, clasping.

CORE Median. Axile or Abaxile.

CELLS Round to roundish-ovate.

SEEDS Obtuse to acute. Large. Numerous. Regular.

LEAVES Medium to small. Acute. Bluntly serrate. Medium thick. Flat not undulating. Slightly upward-folding. Mid grey-green. Fairly downy.

POLLINATION GROUP 3

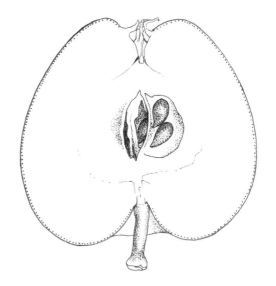

BLENHEIM ORANGE

The Blenheim Orange is one of the lovliest apples of all with its dry distinctive flavour. It was found at Woodstock near Blenheim in Oxfordshire (UK) in about 1740. It is recorded that a countryman named Kempster planted the original kernel and the apple, known locally as Kempster's Pippin, began to be catalogued in about 1818. It received the Banksian Silver Medal in 1820 and thereafter spread through England to Europe and America. The distinctive Classical Blenheim has descended from Kempster's original tree but many clones exist and it seems to be the Broad-Eyed Blenheim that is most common. It is a dual-purpose triploid apple with vigorous growth that requires a dwarfing roostock for restricted space or form. It is a partial tip-bearer and fairly resistant to mildew. The trees bear shyly when young but improve with age. The season is November to January, picking time early October. The flesh is creamy-white, slightly coarse-textured, firm but tender and crisp.

SIZE Large 83 x 70mm (3^{1}/4 x 2^{3}/4").
SHAPE Flat-round. Flattened base and apex. Often lop-sided. Faint well-rounded ribs. Slightly five-crowned at apex. Fairly regular.
SKIN Dull yellow-green becoming yellow. Slightly to half flushed with speckled dull orange to orange-red, fading to orange-yellow at the edges. Indistinct short broad stripes of crimson. Speckled and dotted with fine grey-brown russet dots. Skin smooth and dry becoming greasy.
STALK Stout (4mm) and fairly short (10–14mm). Within cavity or protruding slightly beyond.
CAVITY Fairly deep and wide. Sometimes lipped. Lined with green-ochre and fine scaly light brown russet which straggles over base.
EYE Large. Wide open. Sepals well separated at base, erect with tips reflexed or broken off. Not downy.
BASIN Wide. Quite deep. Ribbed. Slightly russetted.
TUBE Wide. Slightly funnel-shaped.
STAMENS Median.
CORE LINE Basal. Clasping.
CORE Median. Axile, open.
CELLS Obovate.
SEEDS Acuminate. Long thin and pointed.
LEAVES Fairly large. Broadly ovate or broadly acute. Finely and sharply serrate. Medium thick. Flat. Dark green. Slightly downward-hanging. Undersides downy.
POLLINATION GROUP 3. Triploid.

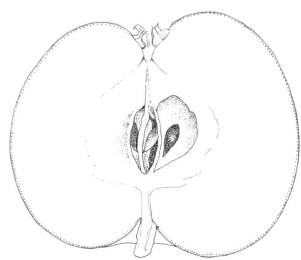

JONAGOLD

This large late season dessert apple is a fairly new introduction and originated from the New York State Agricultural Experiment Station in Geneva, New York. It was raised in 1943 from Golden Delicious x Jonathan. It first fruited in 1953, was introduced in 1968 and received the Award of Merit from the RHS in 1987. The fruits are often very poorly coloured and there are a number of coloured sports, for example Red Jonagold. Trees are triploid, very vigorous, wide-spreading and produce spurs very freely. They are not suitable for the small garden or growing in a restricted form unless grafted on to a dwarfing rootstock. The season is November to February, picking time mid October. The fruits are sweet with a good rich flavour and the flesh is creamy-white, firm, fine-textured and juicy. The skin is a little tough. They have no aroma.

SIZE Large 77 x 64mm (3 x 2^1/$_2$″) or larger on young trees.
SHAPE Round. Base flattened, shoulders often rounded. Flattened at apex. Hint of ribs more noticeable and angular towards apex. Slightly five-crowned at apex. Symmetrical or lop-sided. Regular.
SKIN Light yellow-green becoming more yellow. Slightly to half flushed and mottled bright red with short broken stripes. Occasional small patches of grey-brown russet. Lenticels fairly conspicuous grey-brown russet dots which look raised. Surface dry and bumpy.
STALK Fairly long to long (20–30mm). Fairly slender to medium (2.5–3mm). Protrudes beyond base.
CAVITY Very deep and wide. Partly or more lined with fine ochre or grey russet with a silvery sheen which scatters out over base. Lenticels enter cavity.
EYE Small for size of fruit. Slightly open. Sepals erect and convergent but not touching, with some tips reflexed. Fairly downy.
BASIN Quite wide and deep. Slightly ribbed and puckered. Some cinnamon-brown russet.
TUBE Cone-shaped.
STAMENS Basal.
CORE LINE Basal clasping, often following cell outline.
CORE Median. Axile.
CELLS Obovate.
SEEDS Accuminate. Plump or rather starved. Angular. Straight or slightly curved.
LEAVES Medium size. Acute to broadly acute. Sharply and finely serrate. Medium thick. Flat not undulating. Upward-folding. Light grey-green. Downy underneath.
POLLINATION GROUP 4. Triploid.

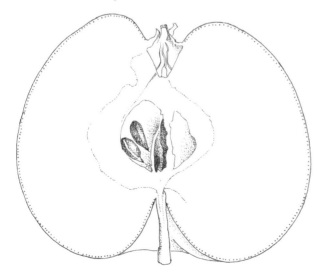

TOWER OF GLAMIS

A very old culinary apple of Scottish origin. The history of this apple appears to be lost but according to the *Herefordshire Pomona* of 1876-85 the variety 'abounds in the orchards of Clydesdale and Carse of Gowrie'. The trees are upright-spreading, compact, moderately vigorous and are spur-bearers. They are hardy and suitable for the north. The season is November to February, picking time mid to end of October. The fruit is sub-acid, rather tasteless but quite pleasant with the addition of some honey. The flesh is white tinged green, firm, crisp, coarse-textured and juicy and cooks to a fluff.

SIZE Medium large 73 x 70mm ($2^7/8$ x $2^3/4$").
SHAPE Round-conical to oblong-conical. Large well rounded or angular ribs, frequently slab-sided. Irregular. Can be four-sided with four prominent ribs. Distinctly four or five crowned at apex.
SKIN Fairly bright grass green turning sulphur yellow. Has a bloomed appearance due to much scarf skin starting at the base and covering half the fruit sometimes extending to apex. Lenticels indistinct pale green or brown russet dots. Skin smooth and slightly greasy.
STALK Medium width (3mm). Fairly short to medium (12–15mm). Level with base.
CAVITY Fairly narrow. Medium depth. Partly or completely covered with golden-brown scaly russet which can streak and scatter over base.
EYE Medium size. Closed or partly open.Sepals broad-based, erect convergent. Very downy.
BASIN Narrow and usually fairly deep. Much puckered and ribbed. Rather pinched looking.
TUBE Long cone or slightly funnel-shaped sometimes reaching almost into core cavity.
STAMENS Median or towards marginal.
CORE LINE Median, but below stamens.
CORE Median. Large and open. Abaxile.
CELLS Obovate or roundish obovate. Can be elliptical. Tufted.
SEEDS Acute. Round. pointed or blunt. Plump. Small and short.
LEAVES Large. Broadly oval. Serrate. Very thick and leathery. Slightly undulating. Dark green. Downward-hanging. Undersides very downy.
POLLINATION GROUP 4

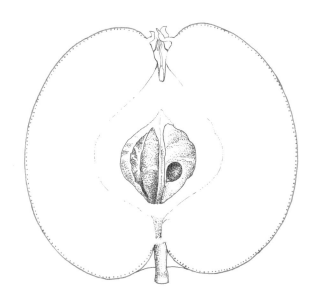

CHIVER'S DELIGHT

An attractive late dessert apple which was raised by Mr. Chivers of Chivers Farms, Histon, Cambridgeshire (UK) in about 1920. The parentage is unknown. The trees are moderately vigorous, upright in habit and produce spurs very freely. They are fairly hardy and suitable for colder areas. The cropping is good and the fruit keeps well. The season is November to January, picking time mid October. The fruits are sweet with a slightly acid bite and an interesting flavour but they are erratic and sometimes flat-tasting. The flesh is creamy-white, firm, fine-textured and juicy. They have a slightly sweet aroma.

SIZE Medium 68 x 58mm (2⅝ x 2¼").
SHAPE Flat-round to round. Base can be wider than apex. Flattened at base and apex. Some fruits lop-sided. Slightly ribbed. Slightly five crowned at apex. Mostly regular.
SKIN Variable. Yellow-green becoming pale golden yellow. Slightly to half dotted and mottled with red with the underlying golden yellow skin showing through. A few short broken rather thin stripes of crimson. Some well coloured fruits have a denser flush of crimson with a little flecking and short thin broken stripes of purplish crimson. There is usually a fairly well defined demarcation line between the flush and the yellow skin. Some small scattered cinnamon-brown russet patches. Lenticels large pale grey-brown, sometimes star-shaped spots.
STALK Slender to medium (2–3mm) Very long (28–35mm). Stalk is thicker where it joins the fruit.
CAVITY Wide to very wide. Medium depth. Partly or more lined with fine grey or scaly grey-brown russet which runs concentrically round cavity and can streak out over base. Can be lipped.
EYE Rather small. Closed or slightly open. Sepals narrow, tapering, erect with tips slightly reflexed or broken off. Stamens visible. Very downy
BASIN Medium depth and narrow. Ribbed and a little beaded. Some specks of cinnamon russet.
TUBE Small. Long cone or slightly funnel-shaped.
STAMENS Median or marginal.
CORE LINE Basal, clasping.
CORE Median. Axile.
CELLS Round or obovate.
SEEDS Acute.Broad. Frequently flattened on one side.
LEAVES Medium size. Acute. Small. Bluntly serrate, Rather thin. Flat. Some upward-folding. Undersides very downy.
POLLINATION GROUP 4

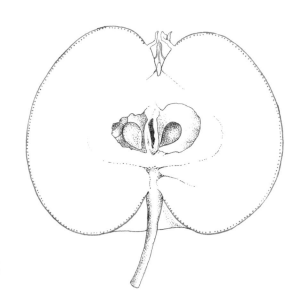

LADY HENNIKER

This is an old English dual purpose apple raised between 1840 and 1850 by John Perkins, gardener at Thornham Hall, Eye in Suffolk, the home of Lord Henniker. The seeds from the apples used for cider were sown and the most promising seedlings selected and grown on. This apple was one of these and became a favourite and carefully preserved. It was finally introduced in 1873 when it received a First Class Certificate from the RHS. The trees are vigorous, upright-spreading and produce spurs freely. They are fairly hardy and the cropping is moderate to good. It is a garden and exhibition variety, season November to January, picking time mid September. The flesh is creamy-white, firm and rather coarse-textured with lovely flavour. When cooked the flesh becomes yellow and breaks up completely.

SIZE Large 73 x 70mm (2⁷⁄₈ x 2³⁄₄").

SHAPE Oblong. Frequently lop-sided. Prominent ribs often with one larger than the rest. Frequently flat or slab-sided with angular ribs and concave between. Irregular. Distinctly five crowned at apex.

SKIN Bright, quite deep yellow-green becoming yellow. A trace to half flushed with thin ochre-orange. Spotted and indistinctly striped with rather thin orange-red. Can have some patchy scarf skin, Variable amounts of fine netted grey-brown russet. Lenticels dark grey-brown or whitish russet dots.

STALK Fairly stout (3.5mm). Short (8–15mm). Within cavity.

CAVITY Deep and wide with stalk well sunk within. Can have one or two pronounced ribs. Some pale brown slightly scaly russet, often underlaid with dark green, which can streak out.

EYE Fairly large. Closed or slightly open. Sepals broad based, sometimes slightly separated, convergent with tips reflexed. Fairly downy.

BASIN Very deep, Narrow. Pronounced ribs. Some fine brown russet.

TUBE Variable. Large cone or funnel-shaped.

STAMENS Median.

CORE LINE Towards basal, clasping, appearing to pull out tube sides.

CORE Slightly distant. Abaxile.

CELLS Obovate. Tufted,

SEEDS Acute. Rather thin. Straight or curved.

LEAVES Fairly large. Broadly oval or broadly acute. Sharply serrate. Medium thick. Flat or slightly undulating. Mid green. Downward-hanging. Very downy.

POLLINATION GROUP 4

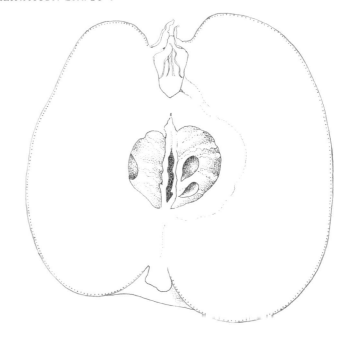

ORLEANS REINETTE

This late dessert apple is thought to have originated in France. Edward Bunyard recalls in 1920 that the variety was 'first described by Knoop in 1776' but does not elucidate. In 1914 it received the Award of Merit from the RHS as Winter Ribston; and again in 1921 as Orleans Reinette. It then came to general notice. The trees are vigorous, upright-spreading and spur-bearers. The season is November to January, picking time mid October.The cropping is rather irregular and they are subject to sudden drop at harvest time and have to be watched. The fruit has a lovely flavour with a suggestion first of sweet oranges following by a nutty flavour after the initial juiciness has gone. The flesh is creamy white, firm, fine-textured and very juicy. The fruits are best stored in polythene bags to avoid shrivelling.

SIZE Medium large 73 x 61mm (2⁷/8 x 2³/8″).
SHAPE Flat-round. Distinctly flattened at base and apex. Symmetrical or slightly lop-sided. Trace of well rounded ribs. Regular.
SKIN Yellowish-green becoming dull golden yellow. Slight to three quarters flushed with dull orange to orange-red with a few rather indistinct short, broken stripes of scarlet. A great deal of the surface is covered with flecks, jagged patches and areas of fine, slightly scaly dull yellow or grey-brown russet. Lenticels conspicuous as jagged or angular russet dots. Skin dry and textured.
STALK Fairly stout to stout (3.5–4mm). Short to medium (10–20mm). Within cavity or beyond.
CAVITY Deep and fairly wide. Sometimes lipped. Green and completely lined with fine grey-brown russet which comes out over base.
EYE Very large. Wide open with stamens clearly visible. Sepals separated at base, flat convergent with tips well reflexed or broken off.
BASIN Wide. Medium depth. Very slightly ribbed but regular. Usually some cinnamon russet which runs concentrically round basin.
TUBE Short, wide, cone-shaped.
STAMENS Median.
CORE LINE Basal, fairly prominent, clasping.
CORE Sessile. Axile, sometimes open.
CELLS Obovate or roundish.
SEEDS Acute. Fairly plump, sometimes angular.
LEAVES Quite large. Broadly oval. Sharply serrate. Fairly thin. Flattish. Very slightly upward-folding. Dark green. Undersides not downy.
POLLINATION GROUP 4. Can be biennial.

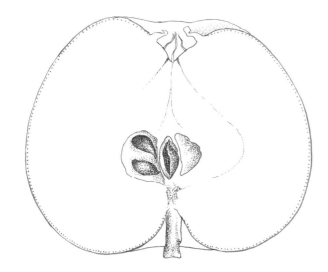

JUPITER

This is a fairly recent dessert apple raised in 1966 at East Malling Research Station in Kent (UK) from Cox's Orange Pippin x Starking. It was named in 1973. It has been planted commercially in the UK but may be superseded by more recent introductions such as Fiesta. It is a late season apple, (November to January), juicy and sweet with a full aromatic flavour. Fruits have a strong and sweet aroma. The skin is a little chewy. The trees are fairly hardy and are suitable for growing in the North. They are vigorous, spreading in habit, spur-bearers and triploid. Picking time is early October. The flesh is creamy-white, rather coarse-textured and juicy.

SIZE Medium 64 x 58mm (2^1/$_2$ x 2^1/$_4$").
SHAPE Conical. Flattened at base and apex. Can be symmetrical or lop-sided. There can be some slight well rounded ribs. Slightly five crowned at apex. Mostly regular.
SKIN Greenish-yellow becoming dull golden yellow. Half to three quarters flushed with orange-red. Very distinct long broad stripes of crimson extending over yellow skin, where they are fainter. A good deal of scarf skin at base and sometimes on cheeks giving the fruit a milky-lilac mottled appearance. Lenticels noticeable as whitish dots often surrounded by an areola of pink. Skin smooth and dry.
STALK Medium to stout (3–4mm) and long (25mm). Frequently fatter where it joins the fruit.
CAVITY Wide. Medium to shallow. Nearly always lipped on one side. Usually lined with greenish-ochre or grey-brown fine scaly russet which can streak out over base.
EYE Medium size. Partly or fully open. Sepals long, broad based, erect convergent with tips reflexed. Fairly downy.
BASIN Fairly wide and deep. Ribbed. There can be a small amount of light brown russet.
TUBE Funnel-shaped.
STAMENS Median.
CORE LINE Median, joining at base of funnel bowl.
CORE Median. Axile.
CELLS Obovate.
SEEDS Acuminate. Rather shrivelled and starved looking.
LEAVES Small. Acute. Serrate. Thick and leathery. Upward-folding and slightly undulating. Mid grey-green. Undersides downy.
POLLINATION GROUP 3. Triploid.

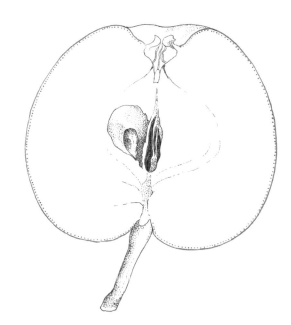

JONATHAN

This late dessert apple originated on a farm at Woodstock, Ulster County, New York where it was raised by Philip Rick and thought to be a seedling of Esopus Spitzenberg open pollinated. Jonathan Hasbrouck brought the tree to the attention of Judge J.Buel who named it after him. It was first recorded in 1826 in an article written by Judge Buel to the New York Horticultural Society. It is grown commercially in many countries, and in the UK. The trees are very susceptible to mildew but show some resistance to scab. They make weak, weeping trees which produce spurs freely. The season is November to January, picking time early October and the cropping is heavy and regular. The fruits are sweet with a week pleasant flavour, somewhat like peardrops. The flesh is white tinged green, tender, fine-textured and fairly juicy. The skin is tough and chewy. Fruits have no aroma.

SIZE Small to medium 67 x 61mm (2^5/8 x 2^3/8").

SHAPE Round to oblong or conical. Flattened at base and apex. Ribs fairly distinct, angular or well rounded. Slightly five crowned at apex. Can be flat-sided. Slightly irregular. Symmetrical.

SKIN Pale whitish-green becoming pale greenish-yellow. Up to three quarters flushed with intense crimson-red, paler red away from the sun. A few indistinct short broken stripes of deep crimson. Surface hammered. Lenticels conspicuous tiny dark grey-brown dots on flush or green dots on yellow skin. A few small grey-brown or gold russet patches. A hair line may be present. Skin smooth, dry and shiny.

STALK Slender (2mm). Short to medium length (8–18mm). Within cavity or protruding beyond.

CAVITY Deep and fairly narrow. Usually partly or completely lined with fine grey-gold russet which can streak out over base.

EYE Small. Closed. Sepals broad based, short, erect convergent, sometimes pinched. Fairly downy.

BASIN Deep and fairly narrow. Usually russet free. Slightly ribbed and sometimes puckered.

TUBE Funnel-shaped or deep cone.

STAMENS Median.

CORE LINE Basal clasping or Median.

CORE Median. Axile, slit.

CELLS Obovate.

SEEDS Numerous. Fairly large. Acute. Longish oval and bluntly pointed. Straight or curved.

LEAVES Small. Acute. Serrate often deeply cut. Thin. Undulating or flat. Upward-folding. Light grey-green. Very downy.

POLLINATION GROUP 3

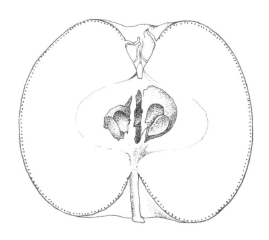

MONARCH

A fairly well known late culinary apple of UK origin that was raised in 1888 by Messrs Seabrook of Chelmsford in Essex from Peasgood Nonsuch x Dumelow's Seedling, and introduced by them in 1918. It is a very good cooker which cooks to a fine yellow fluff with plenty of juice. It is acidic. The trees are fairly hardy and suitable for the West as they show a good resistance to scab. The trees are vigorous, upright-spreading and spur-bearers. The season is November to January, picking time mid September. The cropping is regular and heavy with a biennial tendency. The flesh is white, coarse-textured, tender and juicy. Fruits have almost no aroma.

SIZE Large 80 x 70mm (3^1/8 x 2^3/4") to 73 x 58mm (2^7/8 x 2^1/4").

SHAPE Round-conical to flat-round. Broad flattened base, narrow flattened apex. Symmetrical. No ribs or very slight trace. Slightly five crowned at apex. Can be a hair line at apex. Regular.

SKIN Pale greenish-yellow becoming pale yellow. Slightly to half flushed with grey-red to pinkish-red, which can be sparse and mottled or densely mottled facing the sun. Some indistinct flush-coloured or deep crimson stripes. Lenticels conspicuous green or ochre russet dots on yellow skin and grey russet dots circled with red on flush. Skin smooth becoming greasy if stored.

STALK Medium to fairly stout (3–3.5mm). Fairly short (10–15mm). Often fleshy. Usually level with base or protruding slightly beyond.

CAVITY Deep or medium. Rather narrow. Can be lipped. Some cinnamon-brown russet which can streak and scatter over base.

EYE Quite large. Partly or fully open. Sepals very broad often separated, flat or erect convergent with tips reflexed. Not downy.

BASIN Medium width and depth. Slightly ribbed.

TUBE Quite broad funnel-shaped.

STAMENS Basal.

CORE LINE Basal, clasping.

CORE Median. Axile. Closed or open.

CELLS Ovate.

SEEDS Acute. Fairly plump. Straight or curved.

LEAVES Large. Broadly oval. Crenate. Medium to thin. Slightly undulating. Sightly upwardfolding. Dark green. Slightly downy.

POLLINATION GROUP 4. Biennial.

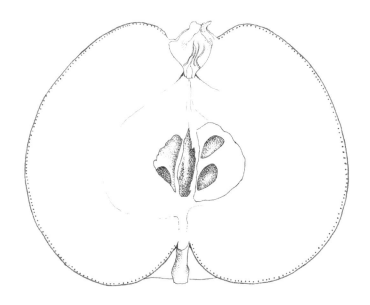

BESS POOL

This old dessert apple was found in a wood in Nottinghamshire (UK) by Bess Pool, the daughter of the village innkeeper. The tree became known and grafts were taken. It was introduced by Mr.J.R.Pearson, a nurseryman of Chilwell, whose grandfather procured the grafts. It was first recorded in 1824.The trees are moderately vigorous, upright-spreading and do not produce spurs very freely being inclined to tip-bearing. It is a useful variety for areas prone to late frosts due to the very late flowering. Trees are shy bearers at first but productivity increases with age. The fruits are rich in flavour with a nice sugar-acid balance though rather dry. The flesh is white occasionally tinged red, firm but tender, slightly coarse-textured and rather dry.

SIZE Medium 67 x 58mm (2⁵/₈ x 2¹/₄").
SHAPE Round-conical. Distinctly flattened at base and apex. Indistinct broad ribs. Symmetrical or lop-sided. Regular.
SKIN Greenish-yellow. Quarter to three quarters flushed with mottled and dotted brownish-red to denser greyed-red. Indistinct broad purple-brown stripes. Partly covered with thin scarf skin giving fruit a milky appearance. Variable amounts of slightly scaly grey-brown russet. Lenticels conspicuous white or grey dots with an areola of greyed-red on yellow skin and purple-red on the flush. Skin dry becoming greasy.
STALK Stout (4mm). Short (10–15mm). Level with base. Can have fleshy knob at spur end.
CAVITY Medium depth and width. Regular. Partly or fully lined with grey-brown sometimes scaly russet which can streak out.
EYE Medium size. Partly open. Sepals erect convergent with tips reflexed. Stamens often visible. Very downy.
BASIN Medium depth and width. Five prominent beads. There can be some grey-brown russet.
TUBE Funnel-shaped or almost cone-shaped.
STAMENS Median.
CORE LINE Median towards basal.
CORE Median. Axile, open or abaxile.
CELLS Ovate.
SEEDS Acute to obtuse. Plump.
LEAVES Fairly large. Acute to narrow acute. Very small serrations. Medium thick. Flat or undulating. Slightly upward-folding. Mid green. Can be downward-hanging. Not downy.
POLLINATION GROUP 6

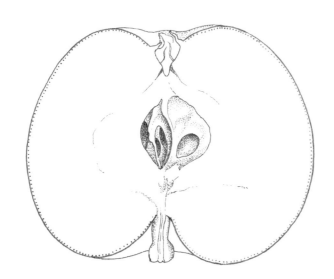

GOLDEN DELICIOUS

Golden Delicious is the most widely planted apple in the main fruit growing areas of the world. It did not originate in France, as one might suppose, but in America, as a chance seedling found by A.H.Mullins of Clay County, West Virginia in 1890. The parentage is uncertain but thought to be from Grimes Golden open pollinated. It was introduced by Stark Brothers in 1914. The trees are fairly frost resistant, moderately vigorous, spreading, and produce spurs very freely but, to succeed, they require reasonably high temperatures and a greater light intensity than is usually found, for example, in England. Fruit grown in cold areas may be small and unattractively russetted. The cropping is good and regular, season November to Feburary, picking time late October. The flavour is sweet with a nice balance of acidity making it refreshing and clear tasting, but rather weak. The flesh is cream, crisp, fine-textured and juicy. Sweet aroma.

SIZE Medium 67 x 64 ($2^5/8$ x $2^1/2$”).
SHAPE Round-conical to oblong. Distinctly ribbed with rather angular ribs more pronounced at apex. Five crowned at apex. Rounded at base. Symmetrical. Fairly regular.
SKIN Pale greenish-yellow becoming yellow. Can be flushed with pale orange. No stripes. Lenticels conspicuous green-brown russet dots larger and more pronounced at base. Some small grey-brown russet patches. Skin smooth and dry.
STALK Slender (2mm) and very long (30–41mm). Protrudes well beyond base.
CAVITY Deep. Narrow. Partly or more lined with fine grey-brown russet which can radiate out.
EYE Medium size. Closed or slightly open. Sepals long,finely pointed, erect convergent with tips partly reflexed. Downy.
BASIN Medium to deep. Medium width. Distinctly ribbed. Some fine cinnamon-brown russet.
TUBE Cone-shaped.
STAMENS Median.
CORE LINE Basal, clasping.
CORE Median. Abaxile.
CELLS Obovate. Quite long and narrow, or elliptical.
SEEDS Acute. Fairly plump. Numerous.
LEAVES Medium size. Long narrow acute. Serrate. Thin. Not undulating. Upward-folding. Mid yellow-green. Slightly downy.
POLLINATION GROUP 4

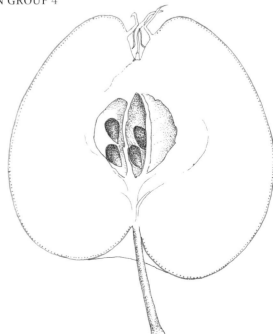

MÈRE DE MÉNAGE

The origin of this old culinary apple is not known. The nomenclature is somewhat confused and it is not the Mère de Ménage of France. It was known in Europe in the late 1700s but it is not known exactly when it was introduced into the UK. The synonyms, Lord Combermere and Combermere infer that someone by this name introduced it. The trees are vigorous, upright-spreading and partial tip-bearers. They crop well. The season is November to February, picking time late September. The fruit is acid with a pleasant flavour, becomes rather dull when cooked and breaks up completely, though not to a fluff. The flesh is greenish-white, firm and crisp, rather coarse-textured and juicy. Fruits have a very slight aroma.

SIZE Large 77 x 67mm (3 x 2⅝″) to very large 86 x 67mm 3⅜ x 2⅝″).
SHAPE Flat-round, some fruits oblong. Distinct ribs, maybe one larger. Very irregular. Frequently lop-sided. Distinctly five crowned at apex, and can have smaller crowns in between.
SKIN Pale yellow-green to pale yellow. Quarter to almost covered with reddish-brown flush, to dark brownish-crimson near the sun. Broad broken stripes of dark purple-crimson. Lenticels conspicuous large pale green or white dots surrounded by green circle. Some scarf skin at base. Skin smooth and dry.
STALK Stout (4–5mm). Short (10–14mm). Usually within cavity.
CAVITY Deep. Variable width. Lined with green-ochre and fine scaly cinnamon russet which can streak on to base.
EYE Large. Partly or fully open. Sepals, sometimes separated, broad based, erect or flat convergent. Some tips reflexed. Very downy.
BASIN Medium depth to shallow. Fairly wide. Distinctly ribbed. Downy.
TUBE Broad and deep cone-shaped.
STAMENS Basal to median.
CORE LINE Almost basal.
CORE Median. Abaxile.
CELLS Roundish obovate. Tufted.
SEEDS Acute. Fairly long, rounded or bluntly pointed. Angular. Plump.
LEAVES Fairly large. Broadly acute. Serrate. Medium thick. Flat or slightly undulating. Dark green. Undersides downy.
POLLINATION GROUP 3

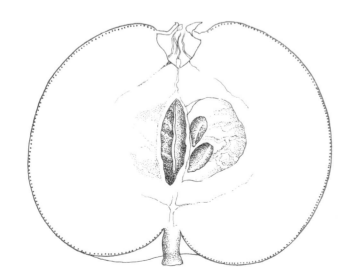

MALLING KENT

This is a late dessert apple which was raised by H.M.Tydeman at East Malling Research Station in Kent (UK) in 1949 from Cox's Orange Pippin x Jonathan. It was named in 1974. The fruit is pleasant, though rather watery, with a good balance of sugar and acid. It has good keeping properties. The trees are moderately vigorous, upright-spreading, and produce spurs freely. The season is November to February, picking time late October. The cropping is heavy. The flesh is creamy-white, slightly coarse-textured, firm, crisp and fairly juicy. The skin is tough. Fruits have a very slight aroma.

SIZE Medium 67 x 67mm (2⁵/8 x 2⁵/8") or 64 x 58 (2¹/2 x 2¹/4").
SHAPE Round-conical to conical. Usually symmetrical. Rounded at base tapering to narrow rather flat apex. No ribs or slight trace. Regular.
SKIN Pale yellow-green to pale greenish-yelow. Up to three quarters flushed with intense crimson. On less colourful fruits flush is more mottled and striped crimson with underlying yellow showing through. Indistinct broad broken stripes of crimson, sometimes over yellow skin, which often enter cavity. Patchy scarf skin at base. Some small patches of fine green-ochre russet. Lenticels inconspicuous green-ochre or grey-brown russet dots. Skin smooth and dry.
STALK Medium to fairly stout (3–3.5mm). Fairly long (18–22mm). Extends beyond base.
CAVITY Medium to deep. Medium width. Cone shaped. Sometimes lipped. Lined with fine grey-ochre russet which can streak over base.
EYE Medium size. Slightly open. Sepals very long and tapering, erect convergent with tips two thirds reflexed. Very downy.
BASIN Wide. Fairly deep. Distinctly ribbed. Some fine brown russet.
TUBE Cone-shaped or slightly funnel-shaped with fleshy protrusion at base entering the tube.
STAMENS Median.
CORE LINE Basal, clasping.
CORE Median. Axile.
CELLS Ovate.
SEEDS Acuminate. Fairly plump. Straight or curved. Light grey-brown.
LEAVES Medium to small. Oval to broadly oval. Bluntly serrate or bi-serrate. Medium thick. Flat not undulating. Some slightly upward-folding. Downward-hanging. Not downy.
POLLINATION GROUP 3

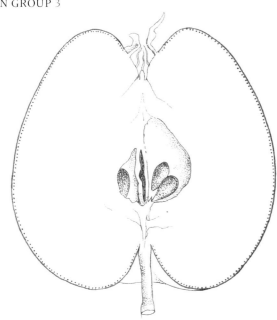

ADAM'S PEARMAIN

This old English dessert apple was orginally brought to notice by Mr.Robert Adams under the name of Norfolk Pippin. Robert Hogg states in *The Fruit Manual* that it was exhibited in Herefordshire, where it orginated, as Hanging Pearmain. The trees are fairly hardy and show some resistance to scab, making them suitable for the West. They are moderately vigorous, wide-spreading and partial tip-bearers. The season is November to March, picking time early to mid October. The cropping is good, with a biennial tendency, and the high quality fruits have the typical rather dry nutty flavour of many of the russets. The flesh is creamy-white, firm, fine-textured, crisp, tender and dry. No aroma.

SIZE Medium 61 x 64mm (2³/8 x 3¹/2") or 70 x 73mm (2³/4 x 2⁷/8").

SHAPE Conical to long-conical. Sometimes waisted. Broad and rounded at base tapering to a flattened apex. Symmetrical or lop-sided. Barely any trace of ribs. Regular.

SKIN Dull pale-green becoming pale golden-yellow. One third to almost completely covered with dull crimson or brighter orangey-red. Broken rather inconspicuous stripes of crimson. Lenticels numerous green-grey or whitish russet dots. Many small pale ochre dots towards apex. Variable amounts of fine ochre or grey-brown russet. Skin dry and slightly rough.

STALK Medium to stout (3–4mm). Short to quite long (13–22mm). Level with base or protruding beyond.

CAVITY Medium width and depth. Sometimes lipped. Lined with greenish-ochre or grey-brown russet speading over base.

EYE Medium size. Slightly to half open. Sepals quite broad, sometimes separated, erect convergent with tips reflexed. Slightly downy.

BASIN Fairly shallow. Medium width. Regular. Slightly ribbed and puckered, sometimes beaded. Occasional areas of ochre-brown russet.

TUBE Funnel or cone-shaped.

STAMENS Median.

CORE LINE Basal or towards median.

CORE Median. Abaxile.

CELLS Variable. Usually obovate, sometimes ovate.

SEEDS Acuminate. Plump. Regular. Mid brown.

LEAVES Small. Oval. Serrate. Rather thin. Flat. Slightly downward-hanging. Mid grey-green. Undersides downy.

POLLINATION GROUP 2. Biennial.

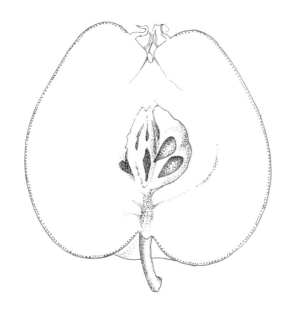

BISMARCK

The origin of this mid to late culinary apple is uncertain. Some think it may have come from Bismarck, Tasmania around the 1870s, others suggest it came from Victoria in Australia or Canterbury in New Zealand. It was named after the German Chancellor, Prince Bismarck. In 1897 it received an RHS First Class Certificate. The trees are very hardy and suitable for colder areas. They are moderately vigorous, spreading in habit and partial tip-bearers. The season is November to February, picking time late September. The cropping is good to heavy. The fruits have a good sub-acid flavour and cook to a yellow fluff. The flesh is white tinged green, fine-textured, crisp and juicy. Aroma Nil.

SIZE Large 77 x 70mm (3 x 2³/₄″).
SHAPE Round-conical. Flattened at base tapering to a narrow flat apex. Five crowned at apex. Frequently lop-sided. Fairly distinct ribs and can be flat-sided. Fairly regular.
SKIN Yellow-green becoming pale greenish-yellow. Half to three quarters flushed brilliant red deepening to crimson. Sometimes a distinct patch of yellow skin in flush when shaded by a leaf. Lenticels conspicuous grey-brown russet dots surrounded by circle of crimson, or tiny pale grey dots. Skin smooth and dry.
STALK Stout (3.5–4mm). Short to medium length (10–18mm).
CAVITY Moderately wide and deep. Green and variably lined with fine scaly dark brown russet, which can radiate out.
EYE Medium size. Closed, occasionally slightly open. Sepals erect convergent and rather pinched in.
BASIN Deep and narrow. Can be rectangular. Ribbed and wrinkled.
TUBE Broad cone-shaped to slightly funnel-shaped.
STAMENS Median.
CORE LINE Median or basal, clasping.
CORE Median. Axile.
CELLS Round to roundish-obovate. Slightly tufted.
SEEDS Sparse. Obtuse to acute. Quite large and plump.
LEAVES Medium size. Acute to broadly acute. Bluntly pointed serrate to crenate. Thin. Flat. Mid green. Downward-hanging. Downy.
POLLINATION GROUP 2. Biennial.

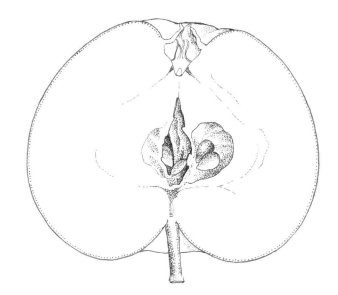

NEWTON WONDER

This is a very large, colourful, late to very late culinary apple. It was raised by Mr. Taylor of King's Newton, Melbourne, Derbyshire (UK), said to be from Dumelow's Seedling x Blenheim Orange. It was introduced by Messrs Pearson & Co in about 1887 and in that year received an RHS First Class Certificate. The trees are too vigorous for the small garden or restricted form unless grafted on to a dwarfing rootstock. They are fairly hardy and suitable for growing in the north and west of the UK, but overlarge fruits are prone to bitter pit. The season is November to March, picking time mid October. The fruits are sub-acid with a good full flavour and cook to a yellow fluff. The flesh is creamy-white, slightly coarse-textured, firm, crisp and juicy. Slight sweet aroma.

SIZE Very large 89–92 x 70mm (3^1/2–3^5/8 x 2^3/4").
SHAPE Flat-round. Symmetrical or lop-sided. Well flattened at base and apex. Can be a slight trace of ribs. Regular.
SKIN Pale yellow-green to greenish-yellow becoming bright golden yellow on ripening. Up to three quarters flushed with brownish-red to very bright red with some golden skin showing through making it look an orange-red from a distance. Distinct broad broken stripes of crimson. Lenticels conspicuous grey-brown or grey-green russet dots. Skin smooth or slightly bumpy becoming greasy.
STALK Stout to very stout (4–6mm). Short (10–p15mm). King fruits fleshy where stalk joins fruit. Level with base or protruding slightly beyond.
CAVITY Fairly shallow and narrow. Can be lipped. Can be a small amount of fine green-ochre russet within cavity.
EYE Large. Fully open. Sepals flat convergent with tips reflexed or broken off. Slightly downy.
BASIN Fairly wide and deep. Regular. Slightly puckered. Some fine grey-brown russet.
TUBE Deep funnel-shaped.
STAMENS Median or towards marginal.
CORE LINE Median.
CORE Median. Axile.
CELLS Fairly small. Round or slightly ovate.
SEEDS Small. Acute. Plump. Straight or curved.
LEAVES Fairly large. Broadly oval or almost round. Serrate or bi-serrate. Medium thick. Flat. Slightly upward-folding. Mid grey-green. Undersides fairly downy.
POLLINATION GROUP 5. Biennial.

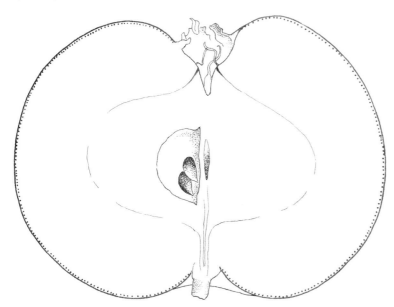

BELLE DE PONTOISE

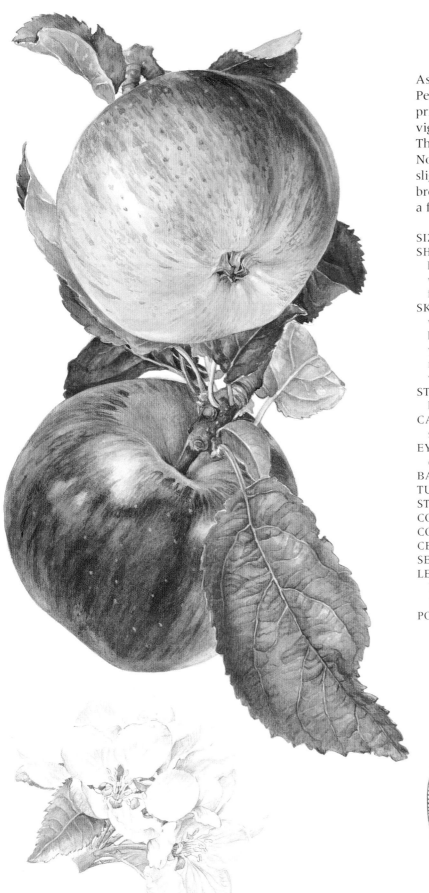

As its name suggests, this apple originated in France, raised by Remy Pere of Pontoise in 1869 from a seed of Emperor Alexander. It is primarily a garden and exhibition fruit. The trees are moderately vigorous, upright-spreading and spur-bearers, producing spurs freely. The cropping is good though it tends to be biennial. The season is November to March, picking time mid October. The flesh is white, slightly coarse-textured, firm and fairly juicy and becomes rather brownish-yellow when cooked. It breaks up completely though not to a fluff. The flavour is sub-acid, pleasant, but rather weak.

SIZE Very large 86 x 64mm (3³/8 x 2¹/2″).

SHAPE Flat-round, sometimes slightly conical. Very broad and flattened at base, flattened at apex. Symmetrical. Irregular with fairly prominent ribs which can be angular. Can be flat-sided or even a bit concave. Occasionally five crowned at apex.

SKIN Pale green becoming pale greenish-yellow. Up to three quarters flushed with dull brownish-red to fairly bright crimson. Broad broken stripes of deep brownish-crimson indistinct on darker flush but can appear over green or yellow skin. Some streaky scarf skin at base. Lenticels conspicuous pale grey russet dots on flush, grey-brown russet dots on green or yellow skin. Skin very smooth and shiny, becoming greasy.

STALK Fairly stout (3.5–4.5mm). Medium length (15–20mm). Protrudes beyond base.

CAVITY Deep and very wide. Can be lipped. Lined with greenish-ochre and some scaly pale brown russet which can spread out over base.

EYE Large. Partly to half open. Sepals broad based, erect convergent or flat convergent with tips fairly erect or broken off.

BASIN Deep and fairly wide. Slightly ribbed, sometimes puckered. Very downy

TUBE Wide and fairly shallow. Cone-shaped.

STAMENS Median or towards basal.

CORE LINE Basal, clasping, sometimes meeting.

CORE Median. Axile.

CELLS Round or slightly obovate.

SEEDS Quite large. Acute. Wide and fairly plump.

LEAVES Large. Broadly acute. Serrate or broadly serrate. Medium thick and leathery. Undulating. Dark yellow-green. Slightly downward-hanging. Undersides downy.

POLLINATION GROUP 3. Biennial.

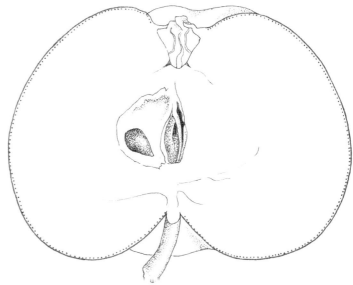

BARNACK BEAUTY

This very late dessert apple was raised in about 1840. It was reported in the *Gardener's Chronicle* of 1899 that the parent tree was still growing in a garden in the village of Barnack, four miles from Stamford near Peterborough in Northants (UK). The apple was introduced in about 1870 by W & J Brown of Stamford and in 1899 received an RHS Award of Merit followed by a First Class Certificate in 1909. The trees are very hardy and will grow in the far north of the UK. They show some resistance to scab. Trees are moderately vigorous, spreading and tip-bearing. The cropping is moderate. The season is December to March, picking time late September. The fruit is sub-acid, juicy and very crunchy with a good refreshing flavour and good balance of sugar and acid.

SIZE Medium 67 x 58mm (2⁵/8 x 2¹/4").
SHAPE Round to oval. Well rounded at base, flattened at apex. Symmetrical or lop-sided. Trace of well rounded ribs. Regular.
SKIN Yellowish-green to pale greenish-yellow becoming yellow. Up to half speckled and dotted with scarlet and the overall appearance is orange-red. Fairly distinct broken stripes of crimson. Some slight patches and specks of grey-brown russet. A hair line may be present. Lenticels conspicuous raised grey-brown russet dots. Surface dry and bumpy.
STALK Medium thick (3mm). Long (20–25mm).
CAVITY Narrow and fairly shallow. Lined with slightly scaly ochre-brown russet that spreads over base. Can be lipped.
EYE Large partly to fully open. Sepals broad based, convergent or erect, sometimes separated at base, with tips reflexed. Stamens usually visible. Slightly downy.
BASIN Wide and fairly shallow. Regular. Slighly puckered. Some golden-brown or scaly grey-brown russet usually around apex spreading into basin.
TUBE Funnel-shaped, sometimes broad or deep.
STAMENS Marginal.
CORE LINE Median joining at funnel neck.
CORE Median. Axile, usually closed, can be open.
CELLS Roundish ovate. Very slightly tufted.
SEEDS Acute or acuminate. Fairly plump. Straight.
LEAVES Medium to small. Acute. Serrate. Rather thin. Flat not undulating. Mid green. Undersides slightly downy.
POLLINATION GROUP 4

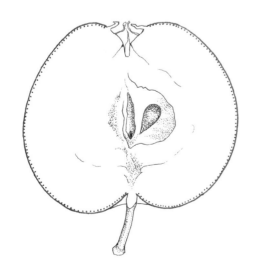

BRAMLEY'S SEEDLING

The Bramley's Seedling was raised in a cottage garden in Church Street, Southwell, Nottinghamshire (UK) by a young lady gardener, Mary Ann Brailsford, from a seed of unknown origin, between 1809 and 1813. In 1848, the cottage was bought by Matthew Bramley and the original tree is still growing in the garden and producing fruit. The apple was named and introduced by Henry Merryweather and first exhibited in 1876. It received a First Class Certificate in 1883. It is the most popular culinary apple in the UK occupying an acreage greater than all other culinary apples combined. The trees are very vigorous, spreading and partial tip-bearers. They are fairly hardy but are susceptible to spring frosts. The cropping is heavy but can be biennial. The season is November to March, picking time mid October. The fruits are acid yet sweet with a good flavour and plenty of juice The flesh is yellowish-white tinged slightly green, firm, coarse-textured and juicy and cooks to a pale cream fluff. There is a red sport named Crimson Bramley.

SIZE Large to very large 89 x 64mm (3$^{1}/_{2}$ x 2$^{1}/_{2}$").
SHAPE Flat-round. Flattened at base and apex. Five crowned at apex. Irregular large ribs, sometimes angular, occasionally flat-sided or concave between ribs. Irregular. Frequently lop-sided.
SKIN Bright green becoming pale greenish-yellow. Can be up to three quarters flushed with pale ochre with some broad broken stripes and dots of greyed-red. Lenticels conspicuous dark grey-brown or green dots. Some scarf-skin at base. Skin smooth and shiny. Becomes greasy on maturing.
STALK Stout (4.5mm). Short (6–10mm). Within cavity or level with base.
CAVITY Medium to deep, generally wide. Partly lined with streaky light grey-brown russet, radiating over shoulder.
EYE Large. Closed or partly open. Sepals broad based, flat convergent, tips reflexed. Very downy.
BASIN Wide. Medium deep. Ribbed, sometimes beaded. Minute russet flecks.
TUBE Slightly funnel-shaped. Almost a cone.
STAMENS Median.
CORE LINE Median, sometimes towards basal.
CORE Median. Axile, open.
CELLS Variable. Generally round, can be obovate or elliptical. Tufted.
SEEDS Acuminate. Rather starved looking.
LEAVES Large. Broadly acute. Finely serrate. Thick and leathery. Undulating and slightly upward-folding. Dark green. Undersides very downy.
POLLINATION GROUP 3. Triploid.

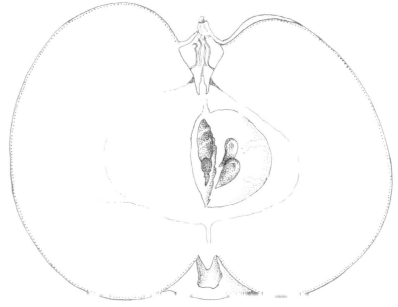

CORNISH GILLIFLOWER

According to Robert Hogg in *The Fruit Manual* the word Gilliflower is derived from the old French word Girofle, signifying a clove and this apple emits a clove-like fragrance when cut. The original tree was found in about 1800 growing in a cottage garden in Truro, Cornwall (UK). Fruits were brought to the Horticultural Society of London in 1913 by Sir Christopher Hawkins and were awarded the Society's Silver medal. It has always been considered a high quality dessert apple possessing a sweet rich flavour. The trees are moderately vigorous, very spreading and tip bearers making them unsuitable for growing in restricted form. The season is November to March, picking time mid October. The cropping is light to moderate. The flesh is yellow tinged green towards core, firm and fine-textured.

SIZE Medium large 70 x 70mm (2³/4 x 2³/4").

SHAPE Oblong to oblong-conical. Very distinct narrow angular ribs. Sometimes flat-sided. Sometimes smaller ribs apparent between the larger ones. Well rounded at base. Distinctly five crowned at apex. Can be lop-sided. Irregular.

SKIN Deep yellow-green to liverish-green becoming yellow. Slightly to half flushed with speckled red with yellow skin showing through. Numerous distinct broad broken stripes of red to purplish-crimson. Some grey-brown netted or dusted russet giving fruits a scruffy appearance. Lenticels very distinct white or green-grey dots. Skin slightly rough or bumpy and dry.

STALK Slender (2–2.5mm) Medium length (18mm). Sometimes fleshy. Protrudes beyond base.

CAVITY Narrow and shallow to medium depth. Irregular. Dark dirty green and partly lined with grey russet.

EYE Medium size. Closed or slightly open. Sepals erect convergent or connivent with tips reflexed. Pinched looking. Downy.

BASIN Narrow and fairly shallow. Distinctly ribbed, sometimes puckered. Some fine cinnamon-brown russet, particularly at apex.

TUBE Cone-shaped or funnel-shaped.

STAMENS Median or marginal.

CORE LINE Median to basal, sometimes two.

CORE Median to slightly distant. Axile or abaxile.

CELLS Roundish obovate. Tufted.

SEEDS Acute. Plump. Dark brown.

LEAVES Rather small. Narrow acute. Bluntly serrate. Thin. Upward-folding. Mid blue-green. Undersides downy.

POLLINATION GROUP 4

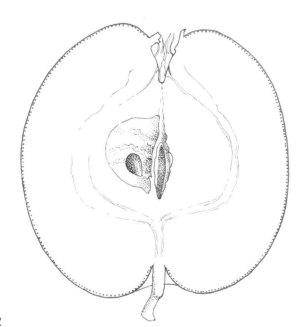

DUMELOW'S SEEDLING

This very late culinary apple was raised by Mr. Dumeller (pronounced Dumelow), a farmer at Shakerstone, Leicestershire (UK), thought to be from seed of the Northern Greening. The original tree was known to be growing in 1800. It was known in that area as Dumelow's Crab but was introduced in 1819 or 1820 by the Turnham Green Nursery as Wellington Apple. This apple used to be widely grown for the mincemeat trade becaue of its good flavour. The trees are moderately vigorous, fairly spreading and produce spurs freely. They are fairly hardy and suitable for the north of the UK. Lenticels are very noticeable on the wood, particularly on young growth. The season is November to March, picking time mid October. The cropping is good. The flesh is whitish, rather coarse-textured, firm, crisp and very juicy. They cook to a very acid pale cream puree.

SIZE Large 77 x 61mm (3 x 2³/₈").
SHAPE Flat-round. Flattened at base and apex. Symmetrical. No ribs. Regular.
SKIN Pale yellowish-green becoming pale yellow. Some fruits quarter to half delicately blushed with pinkish-orange with a few short pinky-red stripes. Lenticels numerous white or pinkish-brown russet dots in a circle of green when not on flush. There can be specks or small patches of fine golden russet. Some scarf skin at base. There can be a hair line. Surface smooth or slightly textured and somewhat pitted. Skin greasy.
STALK Fairly stout (3.5mm). Short (11–14mm). Level with base or protruding slightly beyond.
CAVITY Narrow. Medium depth. Partly lined with fine greenish-gold or grey-brown russet which can scatter over base. Regular. Can be lipped.
EYE Large. Fully open. Sepals fairly long, erect, separated at base with tips reflexed or broken off. Slightly downy.
BASIN Medium to fairly wide and rather shallow. Can have a hint of ribs and sometimes wrinkled and maybe some golden-brown russet.
TUBE Broad funnel-shaped or cone-shaped.
STAMENS Basal.
CORE LINE Basal, clasping.
CORE Median. Axile or abaxile.
CELLS Roundish obovate.
SEEDS Acute. Roundish, plump and often tufted.
LEAVES Medium. Acute. Crenate or bluntly serrate. Medium thick. Flat or slightly undulating, slightly upward-folding. Dark grey-green. Downy.
POLLINATION GROUP 4

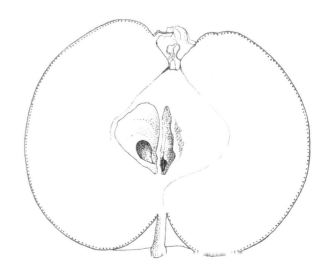

ROSEMARY RUSSET

This is a high quality dessert apple the origin of which is unknown. It was first described in 1831 and was listed in the catalogue of the 1888 Apple and Pear Conference. The trees are moderately vigorous, upright-spreading and produce spurs freely. The season is November to March, picking time late September to early October. The cropping is moderate. It is one of the nicest russets: the fruits have a brisk, sweet yet acid flavour with plenty of juice and are especially good in a warm season, but they are rather too small for commercial use. The flesh is pale creamy-white, firm but not hard, fine textured and juicy. They have a very slight aroma. The trees have a particularly attractive deep pink blossom.

SIZE Medium 64 x 54mm (2¹/2 x 2¹/8″).

SHAPE Conical. Broad and flattened at base tapering to a narrow flattened apex. Slightly five crowned at apex. Sometimes lop-sided. Fairly distinct ribs. Can be flat-sided. Rather irregular.

SKIN Pale greenish-yellow. Can be slightly to half flushed with dull orange or reddish brown. Some indistinct short greyed-red stripes. Partly covered with sparse fine ochre russet with a larger area of pale cinnamon -brown russet at apex. Lenticels conspicuous large grey-brown russet dots. Skin smooth and dry.

STALK Medium thick to stout. (3–3.5mm), occasionally slender (2.5mm). Long (24–28mm). Protrudes well beyond base.

CAVITY Fairly wide and deep. Sometimes lipped. Usually green. Lined with fine grey russet which streaks out over base.

EYE Small to medium. Closed or slightly open. Sepals erect convergent with tips slightly or well reflexed. Can be squashed looking. Downy.

BASIN Medium to narrow. Medium depth. Slightly ribbed and puckered. Up to completely covered with fine grey-brown russet.

TUBE Cone or funnel-shaped.

STAMENS Marginal or median.

CORE LINE Basal clasping or almost basal.

CORE Median. Axile.

CELLS Variable. Ovate or obovate.

SEEDS Acute. Large and numerous. Wide flattish, bluntly pointed.

LEAVES Long narrow acute. Medium size. Serrate. Medium thick. Flat not undulating. Slightly upward-folding. Dark green. Slightly downy.

POLLINATION GROUP 3

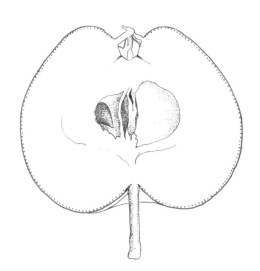

IDARED

This American dual-purpose apple is grown commercially in the UK and often used as a pollinator. It was raised by Leif Verner at The Idaho Agricultural Experiment Station at Moscow, Idaho from Jonathan x Wagener. It was selected in 1935 and introduced in 1942. The trees are fairly hardy, moderately vigorous, upright-spreading and produce spurs very freely. They show some resistance to scab. The season is November to April, picking time end of October to early November. Commercially it is fairly important because of its late keeping properties. The cropping is good. The fruit is rather watery with weak but quite pleasant flavour. The flesh is very white or white tinged green or pink, firm, fairly fine-textured, crisp and juicy. The skin is rather tough. The fruits have no aroma.

SIZE Medium large 70 x 58mm (2³/4 x 2¹/4").
SHAPE Flat-round. Flattened at base and apex. Usually symmetrical. Fairly distinct ribs. Can be flat-sided. Regular.
SKIN Pale greenish-yellow becoming whitish-yellow. Up to three quarters flushed bright crimson-red, pinky-red on less coloured fruits. Indistinct short broken stripes of crimson red to purplish-crimson on well coloured fruits. Usually russet free on cheeks, can have russetted injury patches. Lenticels inconspicuous tiny white russet dots. Skin smooth very shiny and dry.
STALK Fairly slender (2.5mm). Usually long (18–27mm). Protrudes well beyond base.
CAVITY Narrow and deep, often lipped. Usually partly or completely lined with golden-brown russet overlaid with fine grey scaly russet which can streak out over base.
EYE Fairly small. Closed or slightly open. Sepals small, erect convergent or connivent with tips reflexed or broken off. Downy.
BASIN Medium depth. Width medium to narrow. Ribbed and slightly puckered. Usually a dusting or scattered patches of brown russet which can spread over apex.
TUBE Deep slightly funnel-shape or narrow cone.
STAMENS Marginal or slightly towards median.
CORE LINE Median.
CORE Median. Axile.
CELLS Round to roundish obovate.
SEEDS Acuminate to acute. Fairly plump.
LEAVES Medium size. Acute. Serrate. Medium thick. Slightly upward or downward-folding. Mid blue-geen. Undersides downy.
POLLINATION GROUP 2

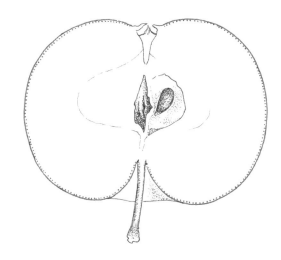

HAMBLEDON DEUX ANS

This old dual-purpose apple came from Hambledon in Hampshire (UK) in about 1750. The trees are very vigorous, upright-spreading and partial tip-bearers and are rather prone to bitter pit. The season is November to April, picking time late September. The cropping is erratic. The fruit has a fair flavour, sweet but with a good acid balance. The flesh is creamy-white, firm, crisp, slightly coarse-textured and dry. It is yellow when cooked and breaks up almost completely but without much juice. The fruits are slightly scented.

SIZE Large 77 x 64mm (3 x 2½").

SHAPE Round-conical to slightly oblong. Flattened at base and apex. Regular or irregular. Some well pronounced ribs. Sometimes flat-sided, often lop-sided.

SKIN Deep grass-green becoming greenish-yellow. Slightly to half flushed with reddish-brown to ochre-brown. Broad broken stripes of dull brownish-purple. A considerable amount of dense or striped scarf skin giving areas a mottled white appearance. Variable amounts of grey-brown russet making the skin feel rough. Lenticels inconspicuous pink or grey-brown russet dots. Skin dry.

STALK Medium to stout (3–4mm). Short (10–15mm). Within cavity or protruding slightly beyond.

CAVITY Medium depth. Fairly wide. Often lipped. Usually dark green and lined with grey-brown slightly scaly russet often spreading out over base.

EYE Smallish. Closed or slightly open. Sepals broad, connivent, sometimes pressed together, with tips reflexed. Very downy.

BASIN Medium width and depth. Irregular. Ribbed and a little beaded. Can be a small amount of grey russet.

TUBE Cone-shaped.

STAMENS Median.

CORE LINE Basal, clasping.

CORE Median. Abaxile, wide open.

CELLS Obovate. Tufted.

SEEDS Acute. Rather starved looking. Long and blunt.

LEAVES Large. Acute. Serrate. Medium thick. Upward-folding, slightly undulating. Dark grey-green. Undersides very downy.

POLLINATION GROUP 3

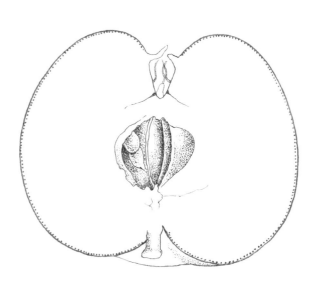

HOLSTEIN

This late dessert apple originated in Germany in about 1918 from Cox's Orange Pippin open pollinated. It was raised or discovered by a teacher named Vahldik in Eutin, Holstein. The fruit is highly aromatic with a Cox-like flavour, sweet with a good balance of acid. It is a delicious apple, one of the best. The trees are very vigorous, wide-spreading and produce spurs freely. They are Triploid. The flesh is creamy-yellow, slightly coarse-textured tender but crisp and juicy. Fruits have a slight aroma, stronger when cut.

SIZE Medium large 73 x 67mm ($2^7/8$ x $2^5/8$ ").
SHAPE Oblong-conical to round-conical. There can be some slight well rounded ribs and can be five crowned at apex. Rounded at base, flattened at apex. Regular, often lop-sided.
SKIN Light yellowsh-green becoming golden-yellow. Up to threequarters flushed. On ripe fruits, flush appears bright orange from a distance but is actually tiny dots and stripes of red over golden yellow skin. Fairly conspicuous thin broken stripes. Small grey russet patches and some patchy scruffy scarf skin at base. Lenticels inconspicuous but raised grey russet dots making surface bumpy. Skin slightly greasy.
STALK Medium to fairly stout (3–3.5mm). Fairly short (9–15mm). Usually protrudes slightly beyond base. Can be set at an angle.
CAVITY Medium to narrow, usually lipped. Fairly shallow. Skin green. Completely lined with fine grey-brown russet, spreading over base.
EYE Large. Wide open. Sepals green, broad, long, erect convergent, sometimes flat-convergent, with tips reflexed. Fairly downy
BASIN Wide and shallow. Slightly ribbed. Usually lined with fine or slightly scaly grey-brown russet which can spread over apex.
TUBE Funnel-shaped.
STAMENS Median, joining at funnel neck.
CORE LINE Median, slightly lower than stamens.
CORE Median. Axile, open.
CELLS Obovate. Tufted.
SEEDS Acute to acuminate. Starved or plump. Broad and straight.
LEAVES Largish. Broadly oval to broadly acute. Serrate. Medium thick. Flat. Margins sometimes curled upwards. Mid blue-green. Undersides very downy.
POLLINATION GROUP 3. Triploid.

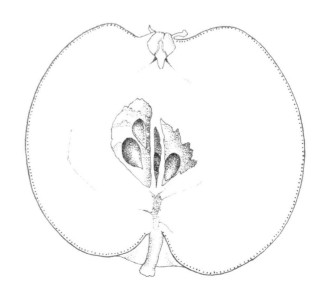

WILLIAM CRUMP

A high quality and attractive dessert apple raised by W Crump of Madresfield Court Gardens, Malvern, Worcestershire (UK). The fruit was exhibited by Mr. Crump on 22nd December 1908 and given the RHS Award of Merit. In January 1910, it was exhibited by Earl Beauchamp of Madresfield and awarded a First Class Certificate. It was introduced by Messrs Rowe of Worcestershire but is not grown commercially in the UK today. The trees are vigorous, upright in habit and spur-bearers. The season is December to February, picking time mid October and the cropping light to moderate. The flesh is cream deepening towards flush, firm, fairly fine-textured, crisp and juicy. The fruit has a rich flavour with a good sugar-acid balance, and is crisp and juicy.

SIZE Medium large 70 x 58mm (2³/4 x 2¹/4").
SHAPE Round-conical. Distinctly flattened at base and apex. Slightly five-crowned at apex. Can be a little waisted near apex. Indistinct ribs with one occasionally more prominent. Usually regular. Fairly symmetrical.
SKIN Yellowish-green becoming yellow. Half to almost completely covered with brownish-crimson flush more red at the extremeties. Indistinct broken stripes of brownish-crimson. Scattered with dots, patches and little scratches of grey russet. Can be some scarf-skin at base. Lenticels conspicuous grey-brown russet dots.Skin smooth and slightly greasy.
STALK Medium thick (3mm). Short to medium length (13mm). Set deep into cavity.
CAVITY Very deep cone. Narrow or broadening out and becoming wide at base. Lined with grey-brown scaly russet.
EYE Medium size. Closed or partly open. Sepals small, erect convergent with tips reflexed. Very downy. Stamens often visible.
BASIN Shallow. Medium width. Trace of ribbing.
TUBE Long narrow funnel-shaped.
STAMENS Almost marginal.
CORE LINE Median.
CORE Median. Axile. Closed or open, or abaxile.
CELLS Ovate.
SEEDS Large. Acute. Broad and plump. Bluntly pointed. Straight.
LEAVES Medium to large. Acute to broadly acute. Serrate. Rather thin. Some undulating otherwise flat. Slightly upward-folding. Mid green. Undersides very downy.
POLLINATION GROUP 5

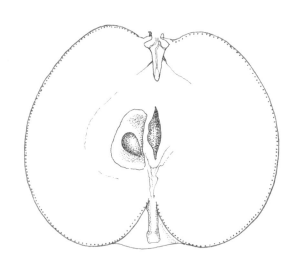

CLAYGATE PEARMAIN

This highly esteemed dessert apple was discovered by John Braddick sometime before 1822. He found it growing in a hedge near his home in Claygate, a hamlet near Thames Ditton in Surrey (UK). On February 19th 1822 John Braddick sent specimens to the meeting of the London Horticultural Society where it was decided that 'it is unquestionably a first rate dessert apple'. In 1901 it received an RHS Award of Merit and in 1921 a First Class Certificate. The trees are moderately vigorous, upright-spreading and partial tip-bearers. The cropping is abundant. The season is December to February, picking time early October. The fruit has a rich almost nutty flavour with a good sugar/acid balance and a very refreshing zest. The flesh is white tinged slightly green, firm, crisp and juicy. No aroma.

SIZE Medium large 70 x 67mm ($2^3/4$ x $2^5/8$").
SHAPE Oblong-conical. Flattened at base and apex. Symmetrical or lop-sided. Trace of well rounded ribs. Regular. Surface bumpy.
SKIN Dull green becoming yellow-green. Up to half covered with dull orange flush deepening to dull red nearest the sun. Short broken stripes of red. Variably covered with slightly scaly grey russet giving it a silverish appearance. Skin dry and slightly rough.
STALK Medium to stout (3–4mm). Short to medium length (10–18.mm). Level with base or protruding beyond.
CAVITY Medium width. Medium to fairly deep. Sometimes lipped. Lined or partly lined with grey sometimes scaly russet.
EYE Large. Open. Sepals broad and long, erect convergent with tips reflexed. Some stamens visible. Not downy.
BASIN Wide and moderately deep. Slightly ribbed. Partly or completely russetted.
TUBE Short funnel-shaped.
STAMENS Median.
CORE LINE Basal, clasping.
CORE Median. Axile.
CELLS Elliptical, sometimes obovate.
SEEDS Acuminate. Very long and sharply pointed. Rather thin.
LEAVES Medium size. Broadly acute. Deeply and sharply serrate. Medium thick. Flat. Slightly upward-folding. Mid green. Very downy.
POLLINATION GROUP 4

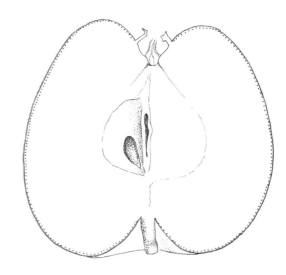

CRISPIN (MUTSU)

Correctly named Mutsu, this apple originated from the Aomori Apple Experiment Station in Japan from Golden Delicious x Indo. It first fruited in 1937.It was re-named and introduced into the UK in 1968 and received an RHS Award of Merit in 1970. The trees are very vigorous, spreading and triploid. They produce spurs freely. The season is December to February, picking time mid October. The cropping is very heavy though with a biennial tendency. The fruits are of good size and it has the advantage of being dual purpose. As a dessert it is fairly sweet with a slight acidity which makes it refreshing. When cooked it stays intact, is pale yellow in colour and has a pleasant though not strong flavour. The flesh is creamy-white tinged slightly green, coarse-textured, firm and fairly juicy, with no aroma.

SIZE Large 77 x 73mm (3 x 2⁷/₈").

SHAPE Oblong. Definitely ribbed and sometimes flat-sided. Ribs can be angular and are more prominent towards apex.Tapers towards five crowned apex. Symmetrical or lop-sided. Fairly regular.

SKIN Yellowish-green with an occasional slight flush of greyed-orange. Lenticels conspicuous green, grey-brown or white dots, larger towards base. Skin smooth and dry.

STALK Fairly slender (2.5mm). Long (22–30mm). Extends well beyond base.

CAVITY Fairly deep and wide, sometimes lipped. Lined with fine golden-brown russet with silvery sheen, which streaks and scatters over base.

EYE Large. Closed or slightly open. Sepals long and tapering, erect convergent, can be slightly separated at base. Fairly downy.

BASIN Medium depth and wide. Ribbed and slightly puckered.

TUBE Cone-shaped.

STAMENS Basal.

CORE LINE Quite pronounced. Basal, clasping.

CORE Median. Axile, open.

CELLS Obovate. Tufted.

SEEDS Acute. Slightly curved. Not plump.

LEAVES Medium to large. Acute to broadly acute. Sharply and deeply serrate. Medium to thin. Flat not undulating. Upward-folding. Mid green. Undersides not downy.

POLLINATION GROUP 3. Triploid. Biennial.

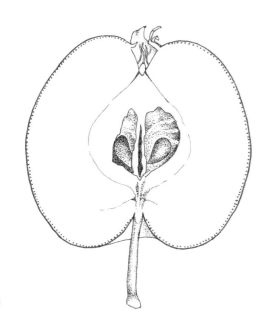

JOHN STANDISH

This apple was first brought to notice in 1921 when it was exhibited at the RHS show by Isaac House & Sons of Bristol (UK) and received a Highly Commended. It is thought to have been raised by John Standish of Ascot in Berkshire in about 1873. In 1922 it received the Award of Merit. It is an attractive scarlet apple the flesh of which is white, very firm, fine-textured, crisp, juicy and fairly acid. There is not a great deal of flavour but what there is is quite pleasant. The skin is fairly chewy. The trees are hardy, vigorous, upright in habit and spur-bearers. The cropping is good. The season is December to February, picking time mid to end October. Fruits have a very slight aroma.

SIZE Medium 64 x 54mm (2^{1}/$_{2}$ x 2^{1}/$_{8}$").
SHAPE Round to round-conical. Rounded at base, slightly flattened at apex. Sometimes a trace of one well-rounded rib. Regular. Symmetrical or lop-sided.
SKIN Very pale greenish-yellow becoming very pale whitish-yellow. Half to three quarters flushed with very bright scarlet red, paler on shaded side. Flush can be deeper crimson on well coloured fruit. Indistinct broken stripes of the crimson which are not very apparent on well coloured fruit. Lenticels conspicuous pale grey or greenish-grey dots on flush or grey-brown russet dots on yellow skin. Small amount of scarf skin at base around cavity. Some small grey-brown russet patches.
STALK Fairly slender to medium (2.5–3mm). Medium to long (15–23mm). Extends beyond base.
CAVITY Shallow to medium depth. Medium width. Partly lined or just a dusting of grey-brown russet that can spread on to shoulder.
EYE Fairly small. Open. Sepals short, erect, convergent with tips reflexed. Very downy.
BASIN Usually very shallow. Sometimes eye is almost standing on summit of apple. Medium width. Slightly ribbed and puckered.
TUBE Funnel-shaped.
STAMENS Median.
CORE LINE Basal clasping, or median towards basal.
CORE Median. Axile, closed or open.
CELLS Round.
SEEDS Obtuse or acute. Round, plump and straight.
LEAVES Medium size. Long acute to long narrow acute. Serrate or bluntly serrate. Medium thick. Flat or slightly undulating. Slightly upward-folding. Mid green. Undersides fairly downy.
POLLINATION GROUP 3

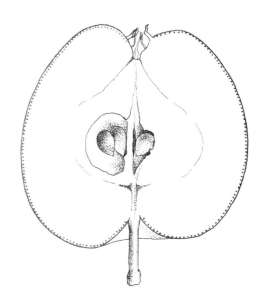

ASHMEAD'S KERNEL

This high quality late dessert apple was raised in Gloucestershire (UK) by Dr Ashmead in about 1700 and until the late 1700s was only really known in West Gloucestershire. The original tree was still in existence during the early 1800s in Dr Ashmead's garden until it was destroyed when the ground was re-allocated for building. It received an RHS Award of Merit in 1969 and a First Class Certificate in 1981. The trees are moderately vigorous, upright-spreading and produce spurs freely. The cropping however is erratic. The season is December to February,picking time mid October. The fruits are sweet yet a little acid and highly aromatic. They are juicy and refreshing. The flesh is yellowish-white tinged slightly green, fine-textured, firm, crisp and juicy. They have no aroma.

SIZE Medium 64 x 54mm (2^{1}/$_{2}$ x 2^{1}/$_{8}$"). Larger on young trees.

SHAPE Flat-round. Frequently lop-sided. Slight trace of well rounded ribs. Can be flat-sided. Well flattened at base and apex. Regular.

SKIN Yellowish-green to pale bright greenish-yellow. Can be partly flushed with ochre-orange to orange-red. Up to almost completely covered with fine slightly scaly cinnamon-brown or grey-brown russet. Lenticels inconspicuous grey-brown or yellow-grey russet dots. Skin very dry.

STALK Short (7–12)mm within cavity, or medium (12–17mm) and protruding slightly beyond. Medium to fairly stout (3–3.5mm).

CAVITY Medium width and depth. Lined with ochre russet and overlaid with specks of deep brown scaly russet, which radiates out.

EYE Medium. Slightly to half open. Sepals erect convergent, tapering, with tips reflexed.Some dark brown russet round eye.

BASIN Medium to shallow. Medium width. Regular. Lined with fine cinnamon and scaly brown russet. Ribbed and puckered.

TUBE Cone-shaped.

STAMENS Marginal

CORE LINE Indistinct, almost basal.

CORE Median, axile.

CELLS Obovate.

SEEDS Acuminate or acute. Fairly plump.

LEAVES Medium size. Broadly acute. Serrate. Medium thick. Flat not undulating. Slightly upward-folding. Mid green. Fairly downy.

POLLINATION GROUP 4

BROWNLEES' RUSSET

This apple was raised and introduced by Mr. William Brownlees, a nurseryman in Hemel Hempstead, Hertfordshire (UK) in about 1848. It is possibly grown on a small scale in order to supply the limited demand for russets, principally in farm shops. The trees are moderately vigorous, upright-spreading and produce spurs very freely. The cropping is good. The season is December to March, picking time mid October. The flesh is greenish-white, fine and firm with a brisk flavour, having a nice balance of sugar and acid, and a pleasant nutty taste. No aroma.

SIZE Medium large 70 x 58mm (2³/4 x 2¹/4″).
SHAPE Flat-round to slightly conical. Fairly distinct well rounded ribs. Flattened to rounded at base with some ribs rather pronounced on shoulder. Frequently lop-sided. Irregular.
SKIN Fairly bright yellow-green to greenish-yellow. Some fruits up to quarter flushed with dull orange-brown to reddish-brown. Skin almost completely covered with greenish ochre and fine grey-brown russet which in parts is overlaid with a fine silvery scaly sheen. A few lenticels are conspicuous pale yellow russet dots. The lenticels enter cavity and are larger towards base. No stripes. Skin dry.
STALK Fairly slender to medium thick (2.5–3mm). Short (8–12mm). Usually level with base or within cavity. Can have fleshy lump on one side.
CAVITY Medium width and depth. Sometimes lipped otherwise regular. Usually completely lined with ochre-brown russet. Some have an underlying dark green skin that radiates out from cavity and over shoulder.
EYE Small to medium. Closed or partly open. Sepals erect convergent with tips reflexed or broken off. Slightly downy.
BASIN Shallow and fairly narrow. Eye sits almost on top of fruit. Slightly ribbed. Green skin next to the eye, otherwise completely russetted.
TUBE Narrow funnel-shaped. (Robert Hogg in *The Fruit Manual* found it to be cone-shaped).
STAMENS Median or almost marginal.
CORE LINE Rather indistinct. Median.
CORE Median. Axile, closed or open.
CELLS Ovate narrowing to a point beneath eye. Tufted.
SEEDS Acuminate or acute. Plump. Dark brown.
LEAVES Small. Long acute to narrow acute. Bluntly serrate. Medium thick. Flat not undulating. Dark blue-green. Undersides downy.
POLLINATION GROUP 3

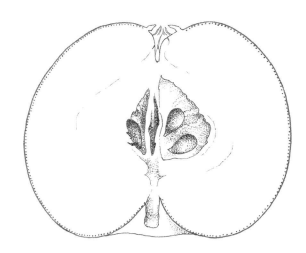

RED DELICIOUS

This apple was disovered on a farm in Wellsberg, Iowa, USA, in about 1880. The farmer, Jesse Hiatt, entered it, under the name Hawkeye, into a competition at Stark Bros show in 1893, and won the prize after which it was renamed Delicious and introduced by Stark Bros in 1895. Later, because the Golden Delicious was also marketed by Stark Bros, it became known as Red Delicious. It requires plenty of sun so it does better in warmer climates but it is one of the most widely planted apple varieties in the world. There are a great many sports. The trees are moderately vigorous, upright-spreading and spur-bearers. The cropping is good. The season is December to March, picking time mid October. The flesh is creamy-white tinged green, fine-textured and firm and very juicy. The flavour is sweet and highly aromatic but the skin is very tough.

SIZE Medium large 70 x 70mm (2³/4 x 2³/4").
SHAPE Oblong to oblong-conical. Flattened at base, rounded shoulders. Very pronounced ribs especially towards apex where they terminate in five high crowns. Can be waisted towards apex. Frequently lop-sided. Irregular.
SKIN Pale dull greenish-yellow usually entirely covered with crimson, greyed-red on shaded side. Fruit is covered with a dense bloom which rubs off and the fruit polishes to an intense shine. Lenticels conspicuous numerous brownish-white dots. No russet. Skin very smooth and dry.
STALK Fairly stout (3.5–4mm),thickening towards attachment to fruit. Medium to fairly long (19–22mm). Protrudes well beyond base. Can be set at an angle.
CAVITY Wide. Medium depth. Sometimes lipped. Frequently some fine grey russet.
EYE Medium size. Slightly open. Sepals narrow and sharply tapered, erect convergent with some tips reflexed.
BASIN Medium width and deep. Prominently ribbed, slightly puckered. Often falls away on one side. Skin in basin usually paler crimson or dull yellow. No russet. Fairly downy.
TUBE Deep funnel-shape.
STAMENS Marginal.
CORE LINE Median.
CORE Median. Axile, open.
CELLS Obovate. Tufted.
SEEDS Acute. Plump. Straight.
LEAVES Medium to small. Acute. Bluntly serrate. Medium to thick. Not undulating. Very slightly upward-folding. Dark green. Undersides fairly downy.
POLLINATION GROUP 3

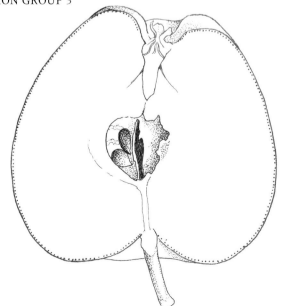

CORNISH AROMATIC

This apple was first brought to notice by Sir Christopher Hawkins in 1813 and it was thought then to have been known in Cornwall for a great many years. It is not sufficiently fertile for commercial planting but has always been a popular garden variety. The trees are vigorous, upright-spreading and make a lot of young growth. They produce spurs freely. The season is December to March, picking time mid October. The flesh is yellowish-white or tinged green, fine-textured, firm and crisp. The taste is sweet, aromatic and slightly spicy with a nice acid bite, but rather dry.

SIZE Medium 64 x 58mm (2^1/2 x 2^1/4").
SHAPE Round-conical to oblong-conical. Usually lop-sided. Distinctly ribbed particularly towards apex. Distinctly five-crowned at apex. Slightly irregular.
SKIN Greenish-yellow becoming soft creamy yellow. Quarter to half flushed with orange-red. Indistinct short broken stripes of crimson. Partly covered with grey-brown russet which can be liberally dusted or in patches. Lenticels very noticeable numerous pale grey russet dots or smaller dark grey dots. Skin dry and rough.
STALK Medium thick (3mm). Medium length (18–21mm). Extends beyond base.
CAVITY Medium width or narrow. Medium to deep. Usually lined with fine ochre-green or grey russet which can streak over base.
EYE Small to medium. Closed or half open. Sepals erect convergent with tips reflexed. Slightly downy.
BASIN Medium depth. Fairly narrow. Distinctly ribbed and some puckering. Can be pinched-looking. Usually russetted with patchy pale cinnamon or grey russet. Sometimes beaded.
TUBE Cone-shaped or slightly funnel-shaped. Small.
STAMENS Median.
CORE LINE Basal, clasping.
CORE Median. Axile.
CELLS Obovate.
SEEDS Acuminate to acute. Fairly plump. Dark brown. Straight.
LEAVES Small. Acute. Bluntly serrate to crenate. Medium thick. Mostly upward-folding, slightly undulating. Mid grey-green. Undersides fairly downy.
POLLINATION GROUP 4

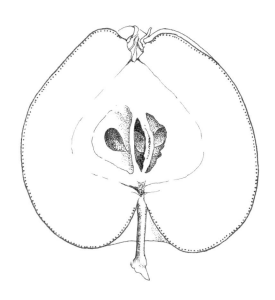

NONPAREIL

This old and highly prized dessert apple has been in existence in England for centuries. It is generally thought that it was brought here from France by a Jesuit in Queen Mary's or Queen Elizabeth's time. It is primarily a garden apple for the connoisseur. The word Nonpareil, meaning a person or thing that is unsurpassed, was applied to a number of apples considered to be of high quality. The trees are rather weak and spreading in growth and produce spurs very freely. The season is December to March, picking time mid October but apples hang on the trees a long time. Cropping is light to moderate. The flesh is greenish-white, firm, fine-textured and juicy. Fruits need plenty of sunshine to develop flavour which can be a little sharp, but in ideal conditions flavour is rich with a good sugar to acid balance. Aroma nil.

SIZE Medium small 58 x 48mm ($2^{1}/4$ x $1^{7}/8$").
SHAPE Flat-round rather conical. Can be slight trace of one or two ribs on some fruits. Base flattened at centre, rounded at shoulders. Flattened at apex. Symmetrical or lop-sided. Regular.
SKIN Pale dull yellowish-green to pale greenish-yellow. Some fruits slightly flushed with greenish-ochre and blotched with pale reddish-brown. No stripes. Partly covered with very fine pale ochre and grey-brown russet. There is a pronounced area of pale grey-brown russet covering apex. Lenticels inconspicuous on cheeks as pale brown russet dots maybe surrounded by prominent patch of greyed-purple. Lenticels more noticeable at base as whitish dots also entering cavity. Skin slightly rough and dry.
STALK Fairly slender to medium (2.5–3mm). Long (18–28mm). Extends well beyond base.
CAVITY Medium width. Medium depth. Regular.Usually lined with fine light grey-brown russet spreading and scattering over base.
EYE Medium size half to completely open. Sepals broad based tapering to a fine point or broken off, erect or convergent but not touching, with tips reflexed. Slightly downy.
BASIN Very shallow. medium width. Slightly ribbed. Mostly lined with fine grey-brown russet.
TUBE Cone or funnel-shaped.
STAMENS Marginal or towards median.
CORE LINE Almost basal.
CORE Median. Axile.
CELLS Ovate or roundish.
SEEDS Obtuse. Fairly plump. Slightly angular.
LEAVES Smallish. Acute. Bluntly serrate. Medium thick. Flat not undulating. Slightly upward-folding. Mid grey-green. Undersides slightly downy.
POLLINATION GROUP 3

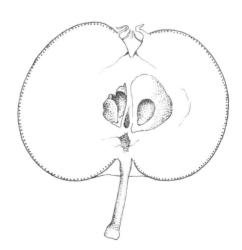

KING'S ACRE PIPPIN

It is stated that this very late dessert apple was introduced in 1899 by King's Acre Nurseries but the exact origin is not specified. It is said to be a Sturmer Pippin and Ribston Pippin cross but which direction is not recorded. It received an RHS Award of Merit in 1897. It has always been highly regarded for its flavour, although its appearance leaves something to be desired. This variety is not suitable for the small garden as it has vigorous growth and makes a large spreading tree. It is a partial tip-bearer. The cropping is slow to begin with but improves when the trees mature. The season is December to March, picking time mid October. The flesh is greenish-yellow, coarse-textured,firm, crisp and juicy and the flavour is richly aromatic and refreshing with a slight acidity. Fruits have a slight richly aromatic scent.

SIZE Medium large 73 x 67mm (2⁷⁄₈ x 2⁵⁄₈").
SHAPE Round-conical sometimes flat-sided and squarish. Usually symmetrical, sometimes lop-sided. Fairly distinct well rounded ribs, one or two can be more pronounced than others. Irregular. Can be slightly five-crowned at apex.
SKIN Dull green becoming dull greenish-yellow. Fruits either not flushed or slightly flushed dull ochre to half flushed brownish-red. Some short broken stripes of dull deep red. All fruits have patches or dusting of fine grey-brown russet, particularly at apex where it is a larger more solid area. Lenticels conspicuous pale grey-brown russet dots. Usually some thin patchy scarf skin at base, sometimes on cheeks. Skin fairly smooth and dry.
STALK Fairly stout (3.5mm).Medium length (17mm). Extends beyond base.
CAVITY Medium to wide. Medium depth. Often lipped. Usually lined with fine grey-brown russet.
EYE Medium size. Closed or partly open. Sepals broad based, erect convergent sometimes not quite touching, with tips reflexed. Fairly downy.
BASIN Medium width and fairly shallow. Ribbed and puckered. Covered with grey-brown russet.
TUBE Funnel-shaped.
STAMENS Median, above core line.
CORE LINE Median almost basal. Rather faint.
CORE Median. Axile.
CELLS Obovate. Slightly tufted.
SEEDS Fairly large. Dark brown.Plump. Acuminate. Curved.
LEAVES Medium size. Acute to broadly acute. Sharply serrate. Thin. Flat. Dark green. Undersides slightly downy.
POLLINATION GROUP 4

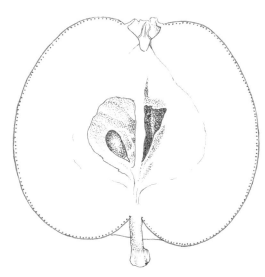

LORD HINDLIP

This very late dessert apple was first exhibited to the fruit committee of the RHS by Mr. Watkins of Pomona Farm, Hereford (UK) in 1896. It was given an Award of Merit and in 1898 received a First Class Certificate when shown by Mr. P.C.M Veitch. The trees are moderately vigorous, upright-spreading and produce spurs fairly freely. The cropping is good. The season is December to March, picking time early to mid October. The flesh is white, tinged pink under the skin, rather coarse-textured, firm and juicy. The fruits have a rich and distinctive vinous flavour with a very slight aroma.

SIZE Medium large 73 x 73mm (2⁷/₈ x 2⁷/₈").
SHAPE Conical to long conical. Very broad base sloping to narrow apex. Base flattened or rounded. Often lop-sided. Farly distinct ribs often with one larger one. Can be angular and flat-sided towards apex. Five crowned at apex.
SKIN Pale greenish-yellow to pale yellow. Up to three quarters flushed with either brownish-orange speckled and dotted with red, or crimson on well coloured fruits. Some short broken broad stripes of crimson. Lenticels numerous and distinct green-grey russet dots. Variable amounts of greyish-ochre or grey-brown russet, sometimes netted. Skin dry.
STALK Medium thick (3mm). Fairly short to medium (15–20mm). Level with base or protruding slightly beyond.
CAVITY Wide and deep. Regular. Lined with fine dirty ochre and scaly dark grey-brown russet which streaks and scatters over base.
EYE Rather small. Closed or very slightly open. Sepals erect convergent or slightly connivent and rather pressed together, with tips reflexed or broken off. Fairly downy.
BASIN Narrow. Medium depth. Slightly ribbed. Patched or netted with fine grey-brown or pale grey russet. Can be slightly beaded.
TUBE Deep narrow cone or deep funnel.
STAMENS Median or almost marginal.
CORE LINE Rather faint. Median.
CORE Distant to very distant. Axile, open.
CELLS Ovate. Can be long and narrow. Slightly tufted.
SEEDS Acuminate or acute. Plump. Straight.
LEAVES Small. Narrow acute. Serrate or bluntly serrate. Medium thick. Very upward-folding. Light grey-green. Undersides very downy.
POLLINATION GROUP 3

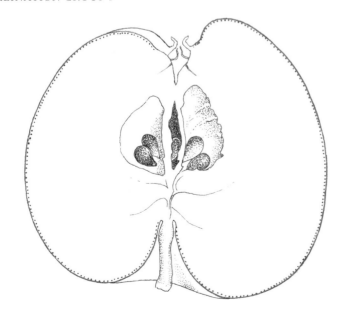

LANE'S PRINCE ALBERT

This late culinary apple is thought to have been raised by Thomas Squire of Berkhamsted (UK) in about 1840 from Russet Nonpareil x Dumelow's Seedling. It was exhibited by H. Lane & Son of Berkhamsted in 1857 and in 1872 received a First Class Certificate from the RHS. The apple was named to celebrate a visit to Berkhamsted by Queen Victoria and Prince Albert. It has been grown commercially in the past but its popularity declined in the 1940s. The trees are hardy and suitable for the north of the UK but are susceptible to mildew. Trees have weak growth and are rather dwarfish making them suitable for the small garden. They are upright then spreading in habit. The season is December to March, picking time early October. The flavour is fair, acidic and a bit watery and it stays intact when cooked. The flesh is greenish-white, fine-textured, firm and juicy. Very slight acid aroma.

SIZE Large 77 x 67 (3 x 2⅝").
SHAPE Round-conical. Often lop-sided. Rather indistinct broad ribs. Slightly five crowned at apex. Base either flat or with slightly prominent ribs. Irregular.
SKIN Bright grass-green changing to light yellow. Some fruits up to one third flushed with greyed-red with broad broken stripes of a deeper greyed-red. Others barely flushed but with greyed-red stripes over a slightly orange skin. Lenticels fairly conspicuous pinky-grey russet or pale ochre dots. No russet. Skin very smooth and shiny.
STALK Fairly slender (2.5mm). Short to fairly short (10–17mm). Within cavity or protruding beyond.
CAVITY Fairly wide and deep. Dark green. Can be partly lined with fine grey-brown russet and some scarf skin which can extend on to base. Some large white oval lenticels in cavity.
EYE Smallish. Closed or slightly open. Sepals small, connivent with tips reflexed. Very downy.
BASIN Fairly deep. Medium width. Ribbed.
TUBE Cone or funnel-shaped.
STAMENS Median.
CORE LINE Basal, clasping.
CORE Median. Abaxile.
CELLS Eliptical or ovate. Tufted.
SEEDS Numerous. Acuminate. Plump. Straight.
LEAVES Large. Acute. Broadly serrate. Medium thick. Mostly flat. Mid green. Undersides slightly downy.
POLLINATION GROUP 3

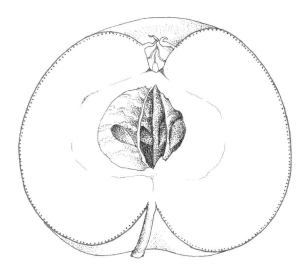

CRAWLEY BEAUTY

This apple was discoveredby Mr. Cheal, a nurseryman of Crawley in Sussex (UK). He found it growing in a cottage garden in Sussex in about 1870. He introduced it in 1906 and it was grown in parts of Sussex earlier last century. The origin of this apple is not certain: one theory is that it is an American cultivar named Goldhanger, but it has also been found to be identical with the French cultivar Nouvelle France. It is a useful very late culinary variety to grow in colder areas because it flowers late and is very hardy. The trees are moderately vigorous, spreading, and produce spurs freely. The cropping is good. The season is December to March, picking time mid October. The flesh is greenish-white, firm, slightly coarse-textured and not very juicy. The flavour is fair. Fruits have no aroma.

SIZE Medium small 58 x 45mm (2¼ x 1¾").
SHAPE Flat-round to round. Distinctly flattened at base and apex. Usually symmetrical, occasionally lop-sided. No ribs. Regular.
SKIN Fairly bright yellow-green becoming pale yellow. Quarter to three quarters flushed with brownish-red tinged crimson nearest the sun, greenish-ochre on shaded side. Fairly distinct stripes of brownish-crimson, except on highly coloured flush where they blend with the flush colour. Lenticels conspicuous whitish dots. A small amount of thin scarf skin at base or in cavity. Usually no russet. Skin very smooth, shiny and greasy.
STALK Medium thick (3mm). Medium length (15–20mm). Protrudes well beyond base.
CAVITY Medium width and depth. Regular. Usually some fine golden russet in cavity.
EYE Medium. Half to fully open. Sepals short, broad-based, separated when fully open, erect convergent with some tips reflexed. Fairly downy.
BASIN Wide. Medium depth. Usually regular with no ribs, occasional puckering. Occasional small amounts of cinnamon russet.
TUBE Slightly funnel-shaped.
STAMENS Median.
CORE LINE Median or towards basal.
CORE Median. Abaxile.
CELLS Obovate.
SEEDS Fairly large. Obtuse. Broad.
LEAVES Small. Acute. Serrate. Medium to thin. Flat not undulating. Slightly upward-folding. Mid green. Undersides very downy.
POLLINATION GROUP 7

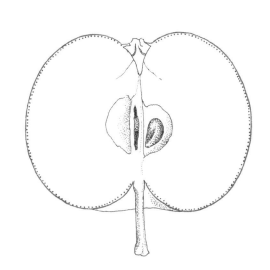

PIXIE

This is a fairly recent introduction. It is a high quality, small, late dessert apple, raised in 1947 at the National Fruit Trials. The parentage is not recorded, but it is thought it may have been a seedling from Cox's Orange Pippin or Sunset. It received an RHS Award of Merit in 1970 and a First Class Certificate in 1972. It is mainly a garden variety being too small for large scale commercial use. The fruit is delicious, crisp, juicy, sweet yet refreshing, attractive to look at and not too large. The trees are moderately vigorous, wide-spreading and spur-bearers. The cropping is good. The season is December to March, picking time mid October. The flesh is creamy-white, fine-textured, firm and juicy. Fruits have no aroma but are slightly scented when cut.

SIZE Medium 64 x 51mm (2½ x 2").
SHAPE Flat-round. Flattened at base and apex. No ribs. Usually symmetrical, can be lop-sided. Regular.
SKIN Greenish-yellow. Quarter to three quarters flushed with orange-red, mottled and dotted towards the edges. Short broken stripes of red. Some greenish-ochre or greenish-grey russet dots and patches. Lenticels conspicuous slightly raised pale ochre or grey-brown dots. Can be some patchy scarf skin at base. Skin smooth and dry and slightly bumpy.
STALK Slender to medium (2–3mm). Medium length (20mm). Sometimes fleshy.
CAVITY Medium width. Medium to fairly deep. Regular. Some greenish-ochre sometimes scaly russet within cavity which can streak or scatter over shoulder.
EYE Small. Closed or slightly open. Sepals short or long, convergent with half to three quarters reflexed. Stamens sometimes present. Downy.
BASIN Medium width and fairly shallow. Slightly ribbed, occasionaly beaded. May be some flecks of fine grey russet.
TUBE Small funnel-shaped.
STAMENS Median
CORE LINE Basal, clasping.
CORE Median. Axile.
CELLS Obovate.
SEEDS Acute. Very large for size of apple. Dark brown. Numerous. Straight. Fairly plump.
LEAVES Medium size. Acute to long acute. Bluntly serrate. Thin. Flat or slightly undulating. Upward-folding. light yellow-green. Undersides downy.
POLLINATION GROUP 4

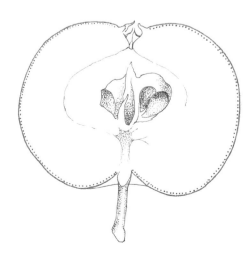

TYDEMAN'S LATE ORANGE

This late dessert apple was raised in 1930 at the East Malling Research Station at Maidstone in Kent (UK) by Mr. H.M.Tydeman from Laxton's Superb x Cox's Orange Pippin. It was introduced in 1949 and has had small scale commercial plantings. It received an RHS Award of Merit in 1965. The fruit has a rich Cox-like flavour with a nice sugar/acid balance. The skin is slightly tough. The cropping is good but trees tend to overcrop with small fruits which should be stored in polythene bags as they tend to shrivel. Trees are vigorous, upright then spreading with long laterals. They produce abundant new growth and spur freely. The season is December to April, picking time mid October.The flesh is creamy-yellow, fine-textured, firm and juicy. The fruits have no aroma.

SIZE Medium small 58 x 54mm (2^{1}/4 x 2^{1}/8"").

SHAPE Conical. Usually symmetrical. Almost no trace of ribs. Flattened at base narrowing to a slightly flattened apex. Regular.

SKIN Pale yellowish-green becoming dull greenish-yellow. Up to three quarters flushed with dull brownish-purple. Indistinct broken stripes of deeper brownish-purple. Patched and slightly netted with fine ochre-grey or brownish-grey russet. Some scarf skin chiefly at or towards base. Lenticels inconspicuous pale or dark grey-brown russet dots. Skin dry and slightly rough.

STALK Fairly slender to fairly stout (2.5–3.5mm). Medium length (15–22mm). Protrudes slightly or well beyond base.

CAVITY Medium width, fairly shallow. Sometimes lipped otherwise regular. Ochre green lined with scaly grey-brown russet which extends over base.

EYE Large. Open. Sepals broad and long, often separated at base, erect with tips slightly reflexed. Stamens present. Not downy.

BASIN Shallow. Medium width. Regular. Slightly puckered. Partly or completely lined with fine pale brown russet.

TUBE Cone-shaped.

STAMENS Median.

CORE LINE Basal, clasping.

CORE Median. Axile, open.

CELLS Roundish ovate.

SEEDS Acuminate to acute. Plump. Regular or curved.

LEAVES Medium size. Acute. Bluntly serrate. Medium thick. Flat not undulating. Some slightly upward-folding. Light grey-green Undersides slightly downy.

POLLINATION GROUP 4

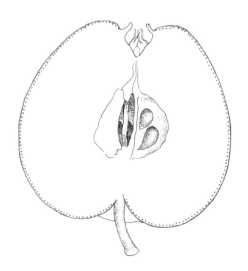

BELLE DE BOSKOOP

This fairly old dual-purpose apple was found by K.J.W.Ottolander at Boskoop, near Gouda in Holland in 1856. It was introduced into the UK and received an RHS Award of Merit in 1897. It is grown commercially in Holland and Germany. It is thought to be a bud sport of Reinette de Montford, but the parentage is not certain. The trees are vigorous, upright-spreading, spur-bearers and triploid. The cropping is good to heavy and it is suitable for growing in the west of the UK. The season is December to April, picking time early October. The flesh is cream tinged green, coarse-textured, firm and rather dry with a rich, slightly sweet and quite acid flavour. It cooks to a golden yellow fluff with a fair, sub-acid flavour, hardly needing additional sugar.

SIZE Medium large 73 x 67mm (2⁷/₈ x 2⁵/₈").
SHAPE Round-conical. Often lop-sided. Trace of ribs with one more prominant. Can be irregular. Flattened at base and apex.
SKIN Light greenish-yellow to light yellow. Up to quarter mottled, dotted and striped with bright red with orange skin showing through making it look more orange-red from a distance. Half to three quarters covered with fine ochre russet, sometimes netted, with base completely covered. Lenticels conspicuous slightly raised grey or green russet dots. Skin dry and hammered.
STALK Stout (4mm). Medium length (17–24mm). Protrudes beyond base.
CAVITY Medium width and deep with stalk well sunk within. Completely lined with fine ochre russet overlaid with brown scaly russet, extending over shoulder and entire base.
EYE Large. Partly to half open. Sepals green, broad based, long and convergent with tips up to half reflexed. Not very downy.
BASIN Medium to wide. Deep. Ribbed and irregular. Partly lined with fine ochre russet.
TUBE Rather wide funnel-shaped.
STAMENS Median.
CORE LINE Median or towards basal.
CORE Median. Abaxile.
CELLS Round, elliptical or slightly obovate.
SEEDS Acute. Long and bluntly pointed. Sparse. Plump or starved.
LEAVES Medium to large. Broadly oval. Finely and sharply serrate. Medium thick. Flat. Mid grey-green. Undersides downy.
POLLINATION GROUP 3. Triploid.

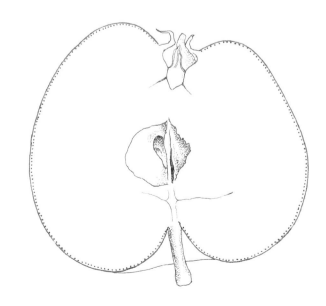

EDWARD VII

A very late culinary apple that was introduced in 1908 by Messrs Rowe of Worcester (UK), said to be from Blenheim Orange x Golden Noble. It was first recorded in 1902 and received an RHS Award of Merit in 1903. It has always had a limited commercial planting in the UK and is a useful garden apple because of its neat upright growth. Trees are rather slow to come into bearing and are only moderate croppers. They flower late and therefore sometimes escape late spring frosts but do not always set well due to lack of pollinators. The trees are hardy and scab resistant and spur-bearers. The season is December to April, picking time mid October. The fruit is acid with a good flavour and cooks to a somewhat red translucent puree. The flesh is creamy-white, rather coarse-textured, firm but tender and fairly juicy. The skin is a little tough. The aroma is nil, but sweetly aromatic on storing.

SIZE Large 83 x 70mm (3¹/4 x 2³/4″).

SHAPE Round to flat-round. Symmetrical or lop-sided. Usually no ribs but can have a slight trace. Regular.

SKIN Fairly bright green becoming light yellow-green and eventually pale yellow. Occasional mottled pale pinkish-brown flush. Some patchy scarf skin at or towards base and surrounding some lenticels. Often a green or dark grey-brown hair line, sometimes two. Lenticels conspicuous grey-brown or whitish russet dots, also some green dots which lie beneath skin. Skin smooth and dry.

STALK Very stout (5mm). Short (6mm). Sometimes fleshy. Usually level with base.

CAVITY Medium width. Very shallow. Sometimes lipped. Lenticels enter cavity as large round or oval white dots. There can be some fine grey-brown russet but usually russet free.

EYE Quite large. Open. Sepals broad based, sometimes separated. Erect convergent with tips reflexed. Slightly downy.

BASIN Medium width. Shallow. Regular. Can be slightly ribbed and puckered.

TUBE Deep funnel-shaped.

STAMENS Median at neck of funnel.

CORE LINE Median.

CORE Median. Axile sometimes open.

CELLS Obovate or elliptical.

SEEDS Acuminate. Fairly plump sometimes angular. Regular or curved.

LEAVES Medium size. Broadly oval to broadly acute. Serrate or bluntly serrate. Medium thick. Slightly undulating. Very dark grey-green. Undersides very downy.

POLLINATION GROUP 6

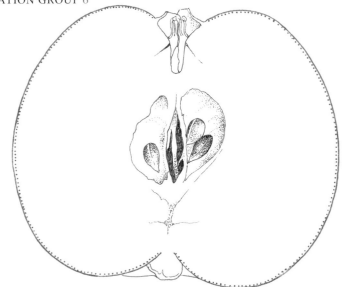

D'ARCY SPICE

This old dessert apple was found in the gardens of The Hall, Tolleshunt d'Arcy, near Colchester, Essex (UK) in about 1785 where it is said many trees of this variety existed at that time. It was always known as D'Arcy Spice or Spice Apple until 1848 when Mr. John Harris, a nurseryman at Broomfield near Chelmsford, sold it under the name of Baddow Pippin. This apple was not grown much outside East Anglia. It requires a hot dry summer to gain the spicy flavour for which it was named. The fruit is richly aromatic, sweet with a good balance of acid, with greenish-white flesh which is firm, fine and juicy. The skin is tough. The cropping is erratic and the fruits shrivel easily unless properly stored. The trees are vigorous, upright-spreading and partial tip-bearers. The season is December to April, picking time late October and early November. The fruits have a very slight aroma.

SIZE Medium 67 x 58mm (2⅝ x 2¼").
SHAPE Oblong. Flattened or rounded at base, flattened at apex. Often lop-sided. Fairly distinct well rounded ribs, often with one larger. Five crowned at apex. Irregular.
SKIN Light yellowish-green becoming pale greenish-yellow. There can be a slight pink to purple-brown flush. Variably covered with finely scaled cinnamon or grey-brown russet. A scruffy looking apple. Lenticels conspicuous slightly raised grey-brown or cinnamon russet dots. Skin very dry.
STALK Medium to stout (3–4mm). Short (12mm). Embedded well and sometimes filling cavity.
CAVITY Fairly narrow and deep. Skin light green with an almost translucent golden hue. Lined with ochre sometimes scaly russet.
EYE Quite large. Partly open. Sepals green, broad based, rather long, convergent with half or just the tip reflexed. Fairly downy.
BASIN Medium width and depth. Ribbed. Partly or completely lined with fine scaly cinnamon russet running concentrically round basin.
TUBE Wide. Cone-shaped.
STAMENS Median or towards basal.
CORE LINE Towards basal or basal clasping. Follows cell outline. Can be a second marginal core line.
CORE Median. Axile.
CELLS Roundish ovate.
SEEDS Acute. Quite large. Flattish. Not plump.
LEAVES Medium size. Acute. Serrate. Medium to thin. Flat. Some downward-folding. Mid green. Slightly downward-hanging. Undersides downy.
POLLINATION GROUP 4

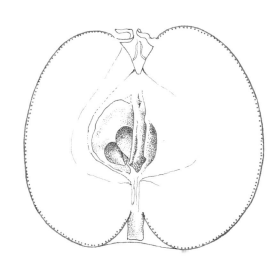

ENCORE

This very late culinary apple was raised by Charles Ross at Welford Park, Newbury, Berkshire (UK) from Warners King x Northern Greening. It was first recorded in 1906 when it received an RHS Award of Merit. It was introduced in 1908 by Messrs Cheal of Crawley and was awarded a First Class Certificate. It is grown on a small scale commercially in the UK. It is a good variety for frosty areas due to its late flowering and it shows some resistance to scab. The trees are moderately vigorous, upright-spreading and spur-bearers. The cropping is moderate. The season is December to April, picking time mid October. The flesh is creamy-white tinged green, rather coarse-textured, soft but firm, sub-acid with a good though not strong flavour. The slices remain intact when cooked. Fruits have a very slightly acid aroma.

SIZE large to very large 83 x 73mm (3^{1}/4 x 2^{7}/8") to 92 x 77mm (3^{5}/8 x 3").

SHAPE Round to oblong. Much flattened at base and apex. Slightly ribbed with broad or angular ribs. Slightly five crowned at apex. Can be a little flat-sided. Occasionally lop-sided. Irregular.

SKIN Bright yellowish-green to greenish-yellow. Indistinctly striped and patched with green under the yellow skin. Can be slightly to half flushed with brownish-crimson or paler greyed brown. Short broken stripes of brownish-crimson. Variable amounts of scarf skin at base. Lenticels fairly distinct white or grey-brown russet dots. Usually russet free or maybe slight amount of fine grey brown russet.

STALK Stout (4mm) and shortish (10–15mm). Set well within cavity. Sometimes fleshy.

CAVITY Medium width and depth. Can be partly lined with fine grey-brown russet.

EYE large. Closed or slightly open. Sepals broad, erect convergent with tips well reflexed. Downy.

BASIN Wide and fairly deep. Ribbed and puckered, sometimes beaded.

TUBE Slightly or definitely funnel-shaped. Deep.

STAMENS Median.

CORE LINE Median.

CORE Median. Abaxile.

CELLS Ovate. Long somewhat lanceolate. Tufted.

SEEDS Acute. Small and numerous.

LEAVES Medium size. Acute. Sharply serrate. Medium thick. Flat not undulating. Mid to light yellow-green. Undersides very downy.

POLLINATION GROUP 4

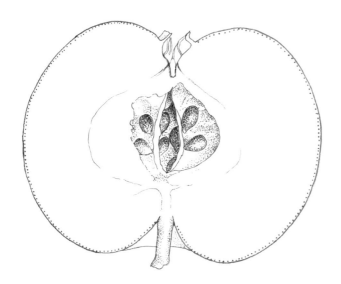

COURT PENDU PLAT

It is suggested that this apple may have originated in Roman times and that it probably came from Europe as it was extensively grown throughout France and Germany. The first record of it is in about 1613. There are a vast number of synonyms attached to this apple, several of which came from the family residences where it was grown, for example Garnon's Pippin, home of Sir John Cotterell, who it is thought may have brought the apple to England. The name is derived from Corps Pendu, referring to the short stalk. It is a useful variety to grow in areas prone to spring frosts due to its late flowering and general hardiness. The tree has rather weak to moderate growth and it makes a smallish upright-spreading tree suitable for a small garden. They are spur-bearers, producing spurs freely and the cropping is good. The season is December to April, picking time mid October. The fruits have a rich aromatic flavour with good sugar/acid balance, and are sweetly scented. The flesh is cream, firm, fine-textured and fairly juicy.

SIZE Medium 61 x 45mm(2³/₈ x 1³/₄").
SHAPE Flat. Very flattened at base and apex. Faint trace of ribs. Symmetrical. Regular.
SKIN Greenish yellow becoming yellow. Up to three quarters flushed with orange to orange-red deepening to scarlet-red on well coloured fruits. Indistinct short broken stripes of crimson slightly more obvious on the orange flushed fruits. Lenticels conspicuous and numerous large green-ochre russet dots. Some thin patches and dusting of grey russet. Skin dry.
STALK Fairly slender to medium (2.5–3mm). Short (10mm). Usually within cavity.
CAVITY Medium width. Fairly deep. Regular. Lined with greyish-ochre russet with fine light brown scaling, which may streak over base.
EYE Large. Open. Sepals short, erect convergent with tips reflexed. Stamens present. Fairly downy.
BASIN Wide and fairly deep. Slightly ribbed. Regular. Some ochre or grey-brown russet dots and dashes lying concentrically round basin.
TUBE Wide funnel-shaped or almost cone-shaped.
STAMENS Median.
CORE LINE Basal or median towards basal.
CORE Median. Axile.
CELLS Rather small. Obovate.
SEEDS Obtuse. Large. Broad. Plump or rather flat.
LEAVES Medium to small. Acute. Broadly, sharply serrate. Medium to thin. Upward-folding. Mid grey-green. Undersides downy.
POLLINATION GROUP 6

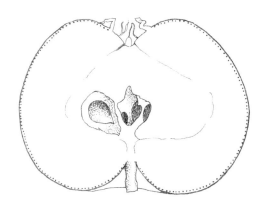

WINSTON

This high quality late dessert apple was raised in the UK in 1920 at Welford in Berkshire by William Pope, from Cox's Orange Pippin x Worcester Pearmain. It was introduced in 1935 as Winter King when it received the Award of Merit from the RHS. In 1944 the name was changed to Winston and in 1951 it received another Award of Merit under the new name. The trees are moderately vigorous, upright-spreading, spur-bearers . The cropping is heavy but the trees tend to overcrop and produce small fruits. The season is December to April, picking time mid October. The fruit has a good aromatic flavour, sweet and slightly acid. The flesh is cream tinged green or pink, fine-textured and fairly juicy. The skin is rather tough.

SIZE Medium small 57 x 54 (2^{1}/4 x 2^{1}/8").
SHAPE Round-conical to oblong-conical. Usually symmetrical, can be slightly lop-sided. Slight trace of ribs. Flattened at apex. Flattened or rounded at base. Regular.
SKIN Dull greenish-yellow. Quarter to three quarters flushed dull brownish-red or darker purplish-red. Fairly distinct narrow or broad broken stripes of dull brownish-crimson, purplish-crimson on darker flush. Lenticels distinct white or grey-brown russet dots. Slight dusting and occasional patches of grey-brown russet.Skin smooth and dry.
STALK Medium to very stout (3–5mm). Short to medium (10–18mm). Level with base or protruding beyond.
CAVITY Medium to shallow. Medium to narrow. Sometimes lipped. Lined with fine ochre-brown russet which can streak over base.
EYE Medium size. Partly to half open. Sepals long, erect-convergent. Downy.
BASIN Medium width and depth. Regular. Slightly ribbed. Some fine grey-brown russet.
TUBE Funnel-shaped. Sometimes with lower portion of funnel filled making core line look basal.
STAMENS Median or towards marginal.
CORE LINE Median (See TUBE).
CORE Median. Axile.
CELLS Small. Obovate sometimes roundish.
SEEDS Acute to acuminate. Plump. Fairly straight.
LEAVES Medium size. Acute. Bluntly serrate. Thick and leathery. Flat not undulating. Sometimes downward-folding. Dark green. Undersides very downy.
POLLINATION GROUP 4

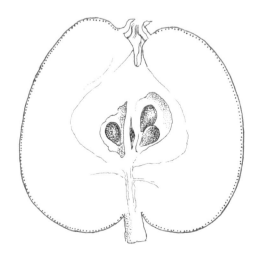

WAGENER

This very late dessert apple is of American origin. It is recorded that in 1791 Mr. George Wheeler brought seeds from Dover, Duchess County, New York, to his new nursery in Yates County. In 1796 Abraham Wagener bought this seedling nursery and by the end of the 1800s, Wagener apple was available from nurseries throughout the United States. It was brought to England in 1910 and received an RHS Award of Merit. The fruit has a rather weak and insipid flavour but is quite refreshing. The flesh is creamy-white, fine-textured, crisp, firm and juicy The skin is tough. It makes a compact upright tree of moderate vigour and it is a spur-bearer. The trees are hardy and suitable for the north and west of the UK. The cropping is heavy with a biennial tendency and it can overcrop with small fruits.

SIZE Medium large 70 x 58mm (2³/₄ x 2¹/₄").
SHAPE Flat-round. Some fairly distinct well-rounded ribs. Symmetrical or lop-sided. Irregular.
SKIN Dull pale greenish-yellow. Quarter to three quarters flushed with bright pinkish-red. Indistinct short broken stripes of crimson. Mottled and streaked with varying amounts of scarf skin chiefly at base. Areas of thin laced and flecked ochre-brown russet, chiefly at apex but sometimes spreading down cheeks. Lenticels inconspicuous small pale whitish dots. Skin smooth and dry.
STALK Fairly slender (2.5mm). Medium length (18–24mm). Protrudes beyond base.
CAVITY Deep. Variable width: wide, or rather compressed with almost ribbed sides. Some ochre or grey-brown russet which can scatter over base.
EYE Medium small. Closed or partly open. Sepals erect convergent, or slightly connivent, with tips slightly reflexed. Downy.
BASIN Medium width and depth. Irregular. Sometimes ribbed. Some lacy ochre-brown russet. Can be beaded.
TUBE Deep funnel-shaped or cone-shaped.
STAMENS Marginal.
CORE LINE Median towards basal. Can be faint.
CORE Median. Axile or abaxile.
CELLS Roundish or roundish obovate.
SEEDS Obtuse. Small to medium. Slightly curved.
LEAVES Medium. Acute to broadly acute. Serrate or bluntly serrate. Rather thin. Slightly undulating and upward-folding. Mid yellow-green. Undersides slightly downy.
POLLINATION GROUP 3

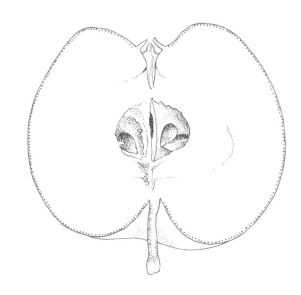

BRAEBURN

This apple needs no introduction, being one of the most important commercially grown dessert apples today and was the first modern apple produced on a large scale to have a good depth of flavour. It originated in New Zealand in the 1950s, thought to be a seedling of Lady Hamilton and possibly Granny Smith. It was named after Braeburn Orchards where it was first grown commercially. The season is January to March, picking time late October. It is not the easiest apple for the UK, requiring a warm climate and long growing season and is not suitable for an organic regime, having little disease resistance.The trees have moderate vigour, are spreading in habit and spur-bearers. The cropping is heavy and trees produce early in life. The flesh is cream tinged green, crisp but not hard, juicy, rather acid but with a good balance of sweetness and a hint of pear-drops. Aroma nil. The variety painted here is Loch Buie, a good variety for growing in the UK.

SIZE Medium 70 x 65mm (2³/4 x 2¹/2").
SHAPE Conical to oblong-conical. Slight trace of some large well rounded ribs which occasionally makes fruit flat-sided.
SKIN Dull lightish-green. Half to almost completely covered with brownish-red to brownish-crimson nearest the sun. Some faint long or broken stripes of slightly darker crimson over flush, more visible on shaded side where they can be long and broad and dull red. Lenticels very conspicuous yellow or pale grey brown dots hardly visible on green skin. Skin smooth and dry.
STALK Medium length (15–22mm). Fairly slender (3mm). Protrudes beyond base.
CAVITY Medium width and fairly deep. Lined with greenish-ochre and varying amounts of fine pale grey-brown russet which can extend over base.
EYE medium size. Partly to fully open. Sepals erect convergent, tips not reflexed or broken off. Stamens often visible.
BASIN Medium to wide. Ribbed. No russet.
TUBE Cone-shaped.
STAMENS Marginal.
CORE LINE Median.
CORE Median.
CELLS Obovate to round. Axile.
SEEDS Mid brown. Acute.
LEAVES Medium to small. Acute. Bluntly serrate or crenate. Fairly flat, some slightly upward-folding. Fairly thick and leathery. Undersides downy.
POLLINATION GROUP 4. Self fertile.

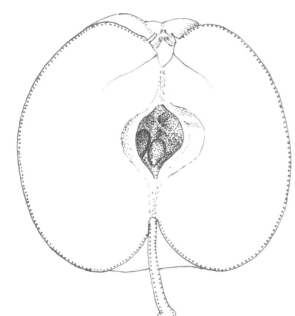

ANNIE ELIZABETH

This fairly old English culinary apple was raised by Samuel Greatorex at Knighton in Leicester in about 1857. In 1866 it received an RHS First Class Certificate and was introduced around that time by Messrs Harrison & Son of Leicester. It is named after the two daughters of Mr. Thomas Harrison, proprietor of the nursery. The trees are fairly hardy and can be grown in all areas of the UK. Trees are moderately vigorous, upright in habit and spur-bearers. The season is December to June, picking time mid October. The cropping is uncertain. The fruits are fairly acid with a good flavour and cook to a fluff. The flesh is creamy-white, rather coarse-textured, crisp, tender and fairly juicy. They have a fairly strong sweet aroma.

SIZE Large 79 x 63mm (3^1/8 x 2^1/2").
SHAPE Roundish-oblong. Slightly flattened at base and apex. Well defined sometimes angular ribs. Irregular. Can be five crowned at apex. Frequently lop-sided.
SKIN Light yellow-green to yellow. Can be quarter to half flecked speckled and striped with greyed-red or densely flushed greyed-red or pinky-red. Numerous short broad stripes of greyed-crimson. Lenticels conspicuous pale brown, green or ochre dots. Some fine scarf skin at base. There can be a hair line. Surface hammered. Skin smooth and fairly greasy.
STALK Medium thick (3mm). Short (9–18mm). Can be fleshy. Level with base or protruding just beyond.
CAVITY Medium depth. Fairly wide. Stalk set well down within. Dark green. Up to completely lined with fine pale brown or ochre russet. Large pale green lenticels in cavity.
EYE Quite large. Closed or partly open Sepals broad, long and tapering, erect, convergent or connivent and rather squashed.Downy.
BASIN Medium width to narrow. Deep. Ribbed and puckered. Some green-brown or ochre russet.
TUBE Deep. Cone-shaped, can enter core cavity.
STAMENS Median or basal.
CORE LINE Median towards basal.
CORE Distant. Abaxile.
CELLS Obovate, sometimes roundish.
SEEDS Acute. Plump. Straight or curved.
LEAVES Large. Broadly acute. Deeply serrate or bi-serrate. Medium thick. Slightly undulating and upward-folding. Dark blue-green. Fairly downy.
POLLINATION GROUP 4

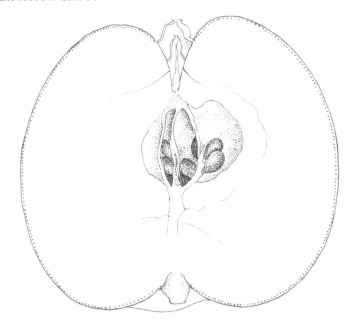

LORD BURGHLEY

This very late dessert apple was raised as a seedling in the gardens of the Marquis of Exeter at Burghley Park, near Stamford (UK). It first fruited in 1834. It was introduced by Mr. Matheson, gardener at Burghley Park, and was distributed in that year by Mr. House of Peterborough. In 1865 it received an RHS First Class Certificate. The trees are moderately vigorous, upright-spreading and have burrs (small rough outgrowths or burls on the trunk and lower branches from which leaf shoots sprout). It is a spur-bearer and produces spurs fairly freely. The season is January to April, picking time mid October. The cropping is fairly good. The flesh is yellowish-white, firm but tender, rather coarse-textured and juicy. The fruits have a rich aromatic flavour, sweet with a nice balance of acidity. They have a moderately strong sweetly scented aroma.

SIZE Small to medium 67 x 61mm (2⁵/8 x 2³/8").

SHAPE Round, slightly conical. Slightly flattened at base and apex. Fairly prominent ribs which can be angular, particularly noticeable towards apex. Slightly five crowned at apex. Irregular.

SKIN Pale yellowish-green becoming pale yellow. Up to three quarters flushed brownish-orange with indistinct stripes and speckles of paler orange-red. Well coloured fruits flushed with brighter orange-red with indistinct thin broken stripes of crimson. Variably patched and dusted with grey-brown russet which looks silvery on a bright flush. Lenticels very noticeable, numerous, large grey or dark brown russet dots. Skin dry and fairly smooth.

STALK Stout (4mm). Medium length (15–20mm). Protrudes beyond base.

CAVITY Medium width. Rather shallow. Sometimes lipped with stalk set at an angle. Partly or completely lined with fine grey-green russet.

EYE Medium size. Partly open. Sepals erect convergent, long and narow with tips reflexed. Downy.

BASIN Medium width, medium depth. Clearly ribbed.

TUBE Deep cone, sometimes almost funnel-shaped.

STAMENS Median sometimes marginal.

CORE LINE Median slightly towards basal.

CORE Median. Axile.

CELLS Obovate.

SEEDS Obtuse. Medium plump. Angular and sometimes flat-sided. Dark brown. Fairly wide.

LEAVES Medium size. Oval. Bluntly serrate. Medium to fairly thick. Flat with some slightly upward-folding. Mid green. Undersides very downy.

POLLINATION GROUP 4

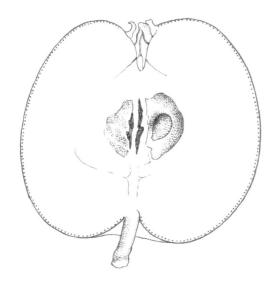

DUKE OF DEVONSHIRE

This well known late to very late dessert apple was raised in 1835 by Mr. Wilson, the gardener at Holker Hall in Lancashire (UK), home of the Duke of Devonshire. It was introduced in about 1875. It used to be grown on a small scale for the markets. It is resistant to scab and is therefore suitable for a damper climate. The trees are moderately vigorous, spreading and spur-bearers. The cropping is good. The season is January to March, picking time early October. The fruits have a creamy-white flesh tinged slightly green, fairly fine texture, and are firm, hard, quite juicy, slightly acidic and refreshing, with a rich flavour. They have almost no aroma.

SIZE Medium 61 x 54mm (2³/8 x 2¹/8").
SHAPE Flat-round to round-conical. Slight trace of ribs. Can be quite flattened at base and apex or rather rounded at base. Symmetrical or lop-sided. Regular.
SKIN Rather dull green becoming pale greenish-yellow and eventually yellow. Partly covered with patchy and netted fine grey-brown russet usually concentrated around the apex and spreading over cheeks. Lenticels conspicuous large, raised, grey russet dots. Skin very dry and slightly textured.
STALK Stout to very stout (4–5mm). Very short (5mm). Well embedded within cavity.
CAVITY Narrow and shallow. Skin green or ochre, lined with fine grey-brown russet, sometimes scaly, which streaks and scatters over base.
EYE Medium size. Slightly to half open. Sepals broad based, short and rather flat or slightly erect convergent with tips reflexed. Fairly downy.
BASIN Shallow. Medium width. Usually fairly distinctly ribbed and puckered. Lined with grey-brown russet with some green skin visible around the eye.
TUBE Cone-shaped.
STAMENS Median towards marginal.
CORE LINE Median or basal, appearing to pull out the sides of the tube and immediately diverging outwards towards the skin.
CORE Median. Axile.
CELLS Obovate.
SEEDS Numerous. Fairly plump. Acute.
LEAVES Medium size. Acute. Bluntly serrate. Medium thick. Slightly undulating and slightly upward-folding. Mid grey-green. Undersides very downy.
POLLINATION GROUP 4

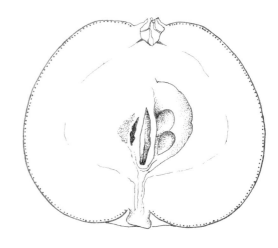

STURMER PIPPIN

This is a fairly old, very late dessert apple which was raised in the UK by a nurseryman named Dillistone at Sturmer near Haverhill in Suffolk. It was recorded in the *Gardeners Chronical* of 1847 that a plant of it was presented to the Horticultural Society by Mr. Dillistone in 1827 and that it was listed as a first rate variety in the Society's Catalogue of Fruits. The parentage is believed to be Ribston Pippin x Nonpareil. It is a high quality apple but requires a warm season to reach its full potential. The fruit has a rich flavour, good and juicy and the cropping is good. The season is January to April, picking time mid to end November. The trees are moderately vigorous, compact and produce spurs very freely.

SIZE Medium 64 x 54mm (2$^{1}/_{2}$ x 2$^{1}/_{8}$").
SHAPE Round-conical to oblong-conical. Frequently lop-sided. Flattened at base and apex. Fairly distinct ribs. Can be flat-sided. Slightly irregular. Five crowned at apex.
SKIN Bright green becoming greenish-yellow. Slightly to half flushed with dull khaki to purplish-brown. Flush stops abruptly without merging. Some slight suggestion of stripes. Frequently russetted with fine cinnamon russet round apex and some small brown russet patches on cheeks. Lenticels distinct on flush as large yellow areolar dots, especially towards base, otherwise white or grey-brown russet dots. Skin smooth and dry.
STALK Fairly slender to stout (2.5–4mm). Fairly short to long (12–25mm). Extends beyond base.
CAVITY Wide and fairly deep. Some greenish-ochre or dark grey-brown russet which can streak and scatter over base.
EYE Fairly small. Closed or slightly open. Sepals broad based and tapering, erect convergent or slightly connivent with tips well reflexed or broken off. Very downy.
BASIN Medium depth and fairly wide. Ribbed. Usually green. Occasional dusting of cinnamon russet.
TUBE Slightly funnel-shaped.
STAMENS Marginal.
CORE LINE Median towards basal.
CORE Slightly sessile. Axile.
CELLS Obovate.
SEEDS Large. Obtuse. Fairly plump. Straight.
LEAVES Medium size. Acute. Bluntly serrate. Medium thick. Flat not undulating. Slightly upward-folding. Mid greey. Undersides very downy.
POLLINATION GROUP 3

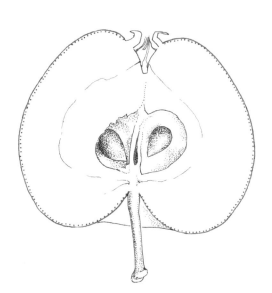

GRANNY SMITH

This apple originated in Australia. It was raised by chance from a seed thrown out by Mrs Thomas Smith of Ryde in New South Wales. The seed is thought to be from a French Crab, open pollinated. The tree was known to be fruiting in 1868. This apple arrived in the UK in 1935. Granny Smith is grown in Australia, South Africa, New Zealand and parts of Europe and North America but it requires a warm climate in order to develop any sugars or flavour and is only suitable as a cooker when grown in the UK. The flavour is insipid but the flesh is hard, coarse-textured, white greenish-white and juicy. The trees are of moderate vigour, upright-spreading and spur-bearers. They make small trees and the cropping is good. The season is January to April, picking time mid October.

SIZE Medium 64 x 61mm ($2^1/2$ x $2^3/8$").
SHAPE Round-conical. Flattened at base with shoulders rounded. Slightly flattened and five crowned at apex. Slightly ribbed. Fairly regular and symmetrical.
SKIN Grass green. Can be slightly flushed with purplish-brown or ochre which can appear rather striped, otherwise no stripes. Lenticels very conspicuous numerous large whitish or pink areolar dots. These dots usually enter the cavity. Skin very smooth and dry.
STALK Slender to fairly slender (2–2.5mm). Medium to long (17–25mm). Protrudes beyond base.
CAVITY Quite deep and cone shaped becoming narrow towards centre with the stalk sunk well within. Some grey-brown russet.
EYE Medium size. Closed or slightly open. Sepals tapering, erect- convergent with some tips reflexed. Very downy.
BASIN Medium width and depth. Usually five definite ribs and some slight puckering. Occasional tiny specks of russet.
TUBE Cone-shaped.
STAMENS Median.
CORE LINE Basal, clasping.
CORE Median. Abaxile or axile.
CELLS Mostly ovate. Sometimes elliptical.
SEEDS Acuminate or acute. Numerous. Straight.
LEAVES Medium size. Rather long acute. Serrate or bluntly serrate. Medium thick. Flat or slightly undulating. Upward-folding. Mid yellow-green. Sightly downward-hanging. Undersides downy.
POLLINATION GROUP 3

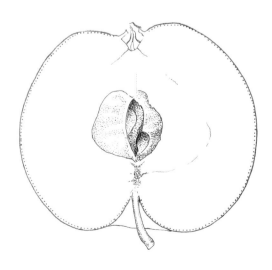

GROWING APPLES

H.A.Baker

The cultivated apple, *Malus pumila domestica,* is not a true species but a hybrid and of a much narrower parentage than was first thought. Its taxonomy has been obscured by a process of hybridization, selection and rejection by man and nature over thousands of years to such an extent that no one today can say what its exact ancestry is. Its evolvement is a continuing process. Modern geneticists, with the objectives of breeding desirable qualities, such as hardiness, heavier cropping, pest and disease resistance, etc. into the apple, are still introducing other *Malus* species into its make-up. But the large fruited apple as we know it today is now thought to have descended from the species *Malus sieversii.*

It is surmised that the apple was first domesticated in the region just south of the Caucasus, as large fruited crab apple hybrids as well as distinct species still exist in primeval forests in that area. It spread from the Caspian Sea across Europe to the Atlantic in pre-historic times. Evidence of such fruits have been found in Neolithic dwellings in Central Europe. Because of its complex parentage, and the wealth of genetic material within its make-up, the apple shows more variability in its progeny than any other major fruit. It does not grow true from seed: every seedling is different. No other fruit has such variability in flavour, texture, shape, colour, or season. There is a cultivar to suit every taste.

Thus if you plant a pip, say from a Cox's Orange Pippin, and it germinates, that seedling will be unique, unnamed and yours to do with as you will. One word of caution to dampen any feelings of optimism about filling the world with wonderful new apples; the chances of raising a seedling better than the many thousands of existing named cultivars is exceedingly remote – a chance in a thousand in fact. Remember, those we have today have come to the top by the hard Darwinian process of rejection and selection carried out by our forefathers and it is still going on.

This is not to say that no good new apple occurs; it does, but not very often and more often than not from professional geneticists based at research institutes. Having said this, two of the most important apples in the United Kingdom at the present time, Cox's Orange Pippin and Bramley's Seedling, were raised by amateurs well over a hundred years ago.

CLIMATE

The apple is a deciduous tree growing in the cool, temperate regions of the world; though, because it has been domesticated by man for such a long time, and because of its great adaptability, there are a few cultivars which will grow and yield quite successfully in the warm, temperate and even in the sub-tropical regions. In the UK and the US it can be grown virtually anywhere given reasonable conditions. Areas subjected to strong winds, exposed coastal regions for example, are unsuitable unless of course ample wind breaks are provided. Strong winds can damage and distort growth, reduce the soil temperature, blow blossom and fruit to the ground and inhibit the movement of those essential pollinating insects. Similarly regions of high altitude present difficulties. The greater the altitude, the cooler the weather and the shorter the growing season. Fruits from such places are smaller, greener, less sweet than those from warmer areas.

In general terms the best apples are grown at altitudes of less than 600 feet, ideally below 400 feet, where the summer temperatures are high and the rainfall low. Commercial apple growing in the UK tends to be concentrated in East Anglia, Kent and the southern counties as far west as the Vale of Evesham and to the borders of Devon.

Professional growers are aiming for perfection. Nevertheless, as already mentioned, the apple is a most adaptable fruit of great diversity and there are few areas in this country where some cultivar or other cannot be grown.

Apples which are especially hardy, and therefore suited to the cooler regions, have been noted in the text.

In similar vein, areas of high rainfall have problems because, unfortunately, the apple can be attacked by fungal diseases associated with wet conditions, in particular apple scab, *Venturia inequalis.* This is the disease which causes black scab-like lesions on the skin of the fruit and in bad attacks, cracking, which in turn allow the ingress of fruit rotting diseases. Scab can also attack the wood and likewise lead to the wood rotting disease apple canker, *Nectria galligena.* Apple scab is more prevalent in the wet areas but is ubiquitous and can occur anywhere whenever warm wet conditions occur. There are a few apples which are resistant, or partially so, to scab and some less prone to canker, and this is noted in their descriptions. Additionally modern science has provided us with fungicidal chemicals capable of preventing or eradicating scab. It is, therefore, possible to grow scab prone apples, such as Cox's Orange Pippin, in areas where otherwise they could not be contemplated. It is appreciated that not every gardener is keen on using chemicals and there is still the problem of how to spray a large tree. Growers reluctant to use chemicals should not be deterred but consider growing the scab resistant cultivars and cooking apples, or be prepared to accept the blemished fruit. Ater all, the skin can be peeled if necessary.

SOIL AND SOIL DRAINAGE

The ideal soil is a slightly acid, about pH6.7, well drained, medium loam eighteen inches or more in depth. However, the apple is tolerant of a wide range. Good drainage is the operative term, as water-logged soil leads to all kinds of problems such as root death, poor growth, low yield, apple canker and possibly the complete loss of the tree. Badly drained ground should be avoided and where this is not possible some kind of drainage system must be installed. Shallow soils over chalk are also unsuitable because of the problems of lime induced chlorosis and the poor water holding capacity of such land. Gardeners on shallow soils should be prepared to dig out the chalk to a depth of not less than fifteen inches by a two foot radius at each planting site and replace it with a good medium to heavy loam. Light soils are acceptable but their moisture retentiveness must be improved by the generous use of organics and with irrigation whenever necessary.

FROST

One of the greatest hazards to successful apple growing is frost at blossom time. Severe frosts at this critical period can destroy the whole or part of the potential crop. When the buds are dormant they are safe, but once they are opened they become increasingly vulnerable. The answer is, in the choice of site, to avoid a frost pocket. Cold air is denser than warm and will therefore gravitate to the lowest point, pushing the warm air upwards as it does so. Areas where cold air collects are called frost pockets. They may be natural, valley bottoms for example, or they may be man- made, a hedge or solid wall impeding the escape of cold air. Remember, if there is any danger of a frost pocket being created, when planting or constructing a hedge or fence, make provision for cold air to flow away. For instance, leave a twelve inch gap at the bottom of a hedge or erect a slatted fence rather than a solid one.

Avoiding a frost pocket is easier said than done: the garden is where the house is. However, all is not lost in such a situation as there are ways and means of mitigating or avoiding the damage, and much depends on the ingenuity of the grower. Covering the trees whenever frost is forecast is the obvious answer, but if the tree is large this is obviously impracticable.

Therefore, whenever new plantings are contemplated, it would be wise to grow trees in restricted form, and on dwarfing rootstocks, so that they will always be small enough to cover if necessary. In areas which are particularly prone, choose cultivars which flower later than normal and therefore stand a better chance of avoiding those late spring frosts.

CHOOSING THE SITE

Select a sunny sheltered position for any new planting. Sunshine and warmth are necessary to ripen the wood, promote the devlepment of fruit buds and give size, colour, flavour and sugars to the apple. The importance of shelter has already been emphasized. Suffice to say, it is essential if a good fruit set and subsequently a satisfactory crop of high quality fruits is to be obtained.

CHOICE OF CULTIVAR AND THE REQUIREMENTS OF CROSS POLLINATION

There are so many lovely apples, as will be seen when looking at the preceding descriptions and pictures, that one is almost spoilt for choice. Remember, when selecting apples for the garden, that no cultivar is truly self-fertile. It will not set a good crop with its own pollen. This means that an apple should not be planted singly but have a partner as pollinator, another but different cultivar which flowers at the same time. Pollinating insects, such as bees, will perform the essential service of cross pollination at blossom time. If there is insufficient room for more than one tree, then plant a family tree, that is, a tree created by the nurseryman with more than one cultivar grafted upon the rootstock. It might consist of two, sometimes up to four, cultivars which are specially chosen both so that cross pollination is achieved and to give a selection of dessert fruits, or perhaps a mixture of dessert and culinary. It is not a wise policy to rely upon a neighbour's tree to pollinate yours, but if there is no other recourse it can sometimes happen that pollination by such means is perfectly adequate. There is a limit to how far a bee will travel in its own pollen and nectar-seeking forays before it returns to the hive, therefore the tree must not be too far away. As a rough guide, cross pollination will be effective up to sixty feet becoming gradually less effective thereafter.

For pollination purposes, apples are grouped together according to their time of flowering.The earliest flowers are in group 1 and the latest in group 7. (See pollination table, page number) In selecting apples for cross pollination, ideally choose those within the same group; an alternative is to choose from those groups immediately adjacent as there will be sufficient overlap. Forecasting the time of flowering cannot be an exact science as there will be variations from year to year, according to the district and the climate.

There is another important point to consider. Most apples are Diploid in their genetic construction, meaning that they can produce fertile pollen in the normal course of reproduction. However, a few apple cultivars are Triploid, which means they have one extra set of chromosomes. Such apples are perfectly fertile on the female side, that is they bear fruit, but they are very poor producers of pollen. Bramley's Seedling is an example. Where these are planted, select two pollinators to pollinate the triploid and each other. Triploids are designated with a T in the pollination table. Incidentally, triploids are usually very vigorous and are best grafted on to one of the dwarfing rootstocks.

There are a few apples which will set a reasonable crop with their own pollen. No real work has been done on the subject, so there is no comprehensive list, and no doubt there will be others from plant breeders, but from observation and experience, it can be said that the following come within this category:

Beauty of Bath	James Grieve
Upton Pyne	Braeburn
Keswick Codlin	Worcester Pearmain
Charles Ross	Lord Derby
Chiver's Delight	Newton Wonder
Crawley Beauty	Red Devil
Ellison's Orange	Red Falstaff
Emneth Early	Saturn
Greensleeves	Scrumptious
Ingrid Marie	Sunset

CHOICE OF TREE FORM

There are two basic categories of apple trees as far as form is concerned. The first are those grown in the open and pruned in the winter; the typical orchard trees in effect. These are the bush, half standard, full standard and spindle bush. The latter is a form now widely used especially by professional growers. The second category are those grown in restricted form and pruned in the summer, such as the cordon, dwarf pyramid, espalier and, more rarely, the fan. These are the forms best suited to the small suburban garden where space is limited, though a small orchard is still a popular feature in the larger garden. Each form is covered in detail below.

Bush, Half Standard and Full Standard
The bush is the most widely grown of the orchard forms. It is an open centred goblet-shaped tree on a short trunk of about two and a half feet. The dwarf bush is a smaller version with a trunk of about one and a half feet, and is grafted on to one of the dwarfing rootstocks.

The half and full standards are merely taller versions of the bush. The basic shape is the same, except that the trunks are longer, and the heads bigger. The half standard has a trunk of about four and a half feet, and the standard, six feet or more. The standards are grafted on to vigorous rootstocks and are in most instances unsuitable for the modern garden because of their size. Large trees can crop very heavily, perhaps too heavily, and their vigorous habit can create problems with pruning, picking and perhaps spraying. Nevertheless, there can be a place for the large tree as a feature or to provide shade.

Spindlebush
Sometimes called the centre leader tree. Its branches start at about fifteen to eighteen inches from the ground. Those at the base are the longest, those at the top are the shortest, so that it is cone-shaped. The spindlebush is usually grafted on to one of the dwarfing rootstocks.

Dwarf Bush

Bush

Half Standard

Standard

Young Spindlebush

Established Spindlebush

The centre leader tree has certain advantages over the open centre bush. The wide angle which the branches make with the central stem are strong and less liable to break under the weight of fruit. Horizontal growth is more fruitful and less vigorous than upright and much of this type of growth is achieved by tying down the young laterals where horizontal growth does not occur naturally. The disadvantage of the spindlebush is that a tall stake is necessary to support it, and the string or weights tying down the laterals make the tree look rather ugly in the garden. Additionally, a strict regime of renewal pruning is necessary to keep it productive and under control. It is only for the enthusiast because if the top growth is allowed to become too dominant it can be very difficult to correct without major tree surgery. All this means is that it is a form only suitable for the skilled fruit grower.

RESTRICTED FORMS

Cordon
This is the most widely planted of the restricted forms and is ideal for the small garden. Being closely spaced, i.e. two and a half to three feet apart, means a goodly number of cultivars can be planted in a relatively small area. Moreover, as apples are not self fertile, the essential requirement of cross pollination is easily covered.

The oblique cordon planted at an angle of forty-five degrees, rather than vertical cordons, is to be preferred. Top growth on a vertical cordon can grow away too strongly at the expense of the lower fruit spurs. Planting at an angle is a happy compromise between growth and fruitfulness and it also gives you extra length of cropping stem relative to height. Cordons are ideal for growing against the wall or fence. Out in the open a post and wire fence is necessary to support them, with the top wire at about six feet.

Espalier
The espalier consists of horizontal arms or tiers, more or less opposite each other, arising from a central vertical stem. It needs more lateral room than the cordon because of its outstretched arms. It can be a low or high form if necessary, depending upon the number of tiers. A two tier espalier, for example, can be kept to a height of between three and four feet, whereas a four tier espalier needs a height of about six and a half feet. Espaliers make attractive boundary markers between one part of the garden and another – around the vegetable plot, for example, or as a hedge between the vegetables and the ornamentals. It is a decorative form – the outstretched arms look handsome in the spring, covered in blossom, and pleasing in the autumn, covered with fruit.

Dwarf Pyramid
This is similar in shape to the spindlebush except that it is not as tall, being kept down to a height of no more than six feet by summer pruning. The dwarf pyramid is always grafted on to one of the very dwarfing rootstocks. It is closely spaced and again ideal for the small garden, but it needs to be planted out in the open and not against fences and walls as with the cordon and espalier. The dwarf pyramid needs support, which can either consist of an individual stake to each tree or by the use of two horizontal wires at eighteen and thirty-six inches to which the trees are tied.

Fan
The fan is an unusual form for the apple, being mainly reserved for stone fruits, which is a pity because it is an attractive shape. Where the garden has walls or fences of seven feet high or more, the fan is an alternative form to the espalier or cordon.

CHOICE OF ROOTSTOCK

Because the apple does not grow true from seed the only way of perpetuating and multiplying a desired cultivar is by vegetative means. It is an intersting thought that of the many thousands of Cox's Orange Pippin trees in existence today they all stem from one single apple seedling raised by Mr. Richard Cox at Colnbrook Lawn, Slough in 1825.

To take a cutting from a selected tree might seem the obvious way of propagation but most apple cultivars do not root readily by this method. For this reason, and others, the apple is propagated by grafting. Scion wood from the chosen cultivar is budded or grafted on to a compatible rootstock. The rootstock, more than any other factor, governs the eventual size of the mature tree, and it is important to make the correct choice, so that the tree or trees are right for the space available and the tree form in which they are to be grown. Most commercial orchards are closely planted and farmed intensively and, sadly perhaps, most modern gardens are small, therefore the rootstocks mainly used by nurserymen today are dwarfing or semi-dwarfing. Incidentally, these stocks have the added advantage of making the tree precocious, so they bear fruit quickly. The principal rootstocks in order of vigour are:

M.27 Extremely dwarfing. Suitable for the dwarf centre leader trees and cordons. Requires a fertile soil and trees need support throughout their lives. It is excellent for vigorous cultivars but unsuitable for weak ones. It is not a stock which will tolerate neglect. Very precocious.

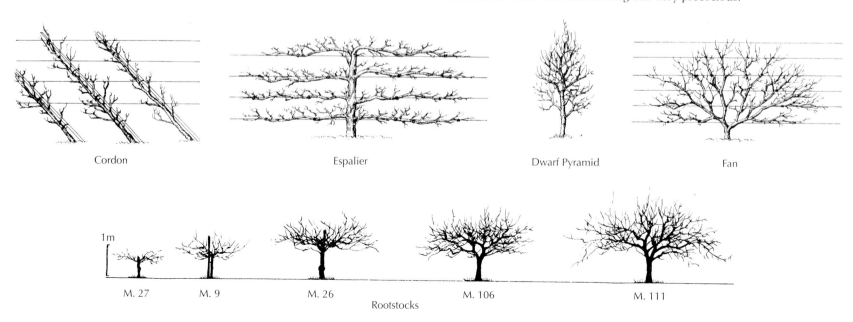

Cordon   Espalier   Dwarf Pyramid   Fan

1m

M. 27   M. 9   M. 26   M. 106   M. 111

Rootstocks

M.9 Very dwarfing. Suitable for dwarf bushes, pyramids, spindle bushes and cordons. It requires a good soil and the trees need staking throughout their lives.

M.26 Dwarfing. A good all-round rootstock suitable for all forms, including the small espalier. Ideal for average soil conditions.

MM.106 Semi-dwarfing. The stock most widely used by nurserymen because it is suitable for most forms and soils, including the lighter ones. Trees out in the open on this stock need staking for the first four or five years.

MM.111 & M.25 Vigorous. Suitable for half and full standards. They could be used for the restricted forms such as the espalier and cordon but only on poor soils otherwise their vigour would become a problem.

Spacing Relative to Tree Form

Spacings given below are not precise but intended as a guide only, because tree size will vary to some extent depending upon the cultivar, environment and the growing conditions.

ORCHARD TREES

Rootstock	Dwarf bush	Bush	Half and Full Standard	Spindlebush
M.27	4' to 6'	–	–	4' to 6'
M.9	8' to 10'	–	–	6' to 8'
M.26	–	10' to 12'	–	8' to 10'
MM.106	–	12' to 15'	–	8' to 12'
MM.111 & M.25	–	–	15' to 20'	–

RESTRICTED FORMS

Rootstock	Dwarf Pyramid	Cordon	Espalier	Fan
M.27	4' to 6'	2½'	–	–
M.9	6'	2½'	–	–
M.26	6'	2½'	10'	10'
MM.106	6' to 7'	2½' to 3'	12' to 15'	12' to 15'
MM.111 & M.25	–	–	15' to 18'	15' to 20'

PLANTING IN THE OPEN

Soil Preparation and Planting

The best time to plant is when the trees are dormant: ideally in the autumn immediately after leaf fall in November, or in the early spring in March before the winter is over. It is not a good policy to plant in the middle of winter when the soil is wet, cold or frozen. In the meantime, heel in the trees in a sheltered spot. It is most important that the roots do not dry out or are subjected to frost. If the ground is frozen solid and it is not possible to heel the trees in, put them in a garden shed or somewhere cool. Keep the roots covered so that they do not dry out but unwrap the aerial parts so they are not forced prematurely into growth. Prepare the ground well. Remove all perennial weeds, by the use of chemical weed killers if desired. Where grass land is involved double digging is necessary, or, for large scale apple growing, deep ploughing. Turn the turves upside down. It would be wise to kill the grass first, especially if couch is present , using a weed killer. Where closely spaced trees are to be planted, turn the whole plot over but for a single tree it is sufficient to prepare the land over a radius of about two feet at each planting site.

Medium to heavy fertile land will not need much bulky organics in the planting hole and it will suffice to fork in a bucketful of compost, a handful of bonemeal and four ounces of a balanced fertilizer. Light soils will need more generous treatment, say two bucketfulls of compost plus the other materials mentioned. The bulky organics are essential to improve the moisture retentiveness of the soil and its structure.

As already mentioned, the ideal time to plant is in the dormant season, using bare rooted trees dug up with a substantial root system, the kind supplied by a reputable fruit tree nurseryman. Containerized plants, whilst they have the advantage that they can be planted at any time, even in the growing season, unfortunately often have the root system drastically reduced in order to fit into the container.

Plant the tree to the same depth as it was in the nursery. The soil mark on the stem of the tree will be an indicator. It is important that the union between rootstock and scion is well above soil level, not less than four inches, to avoid the risk of scion rooting. Many nurserymen graft at about twelve inches so there is no danger. Remember, whilst the tree is out of the ground the root system must not be allowed to dry. If necessary the tree should be heeled in in a sheltered spot until wanted and, throughout the planting operations, keep the roots covered either with soil or with a piece of damp hessian.

All apple trees, no matter what their form, should be supported for the first four years at least. Trees on the very dwarfing rootstocks and those grown in restricted form require permanent support. Cordons, espaliers and fans, for example, are tied to canes affixed to wires.

For trees out in the open, measure the size of the root system and dig out a hole slightly larger and deeper. Drive the stake in first and then plant the tree to the stake. Place the tree three inches away from the tree stake to allow for the eventual expansion of the tree trunk. Ensure that there are no branches chafing against the stake on that side. With bush trees it is usual to drive the stake in so that the head is clear of any branches on that side. Replace the soil, and firm as planting proceeds so that it makes good contact with the roots. Finally, secure the tree to the stake with a figure of eight tie. One can buy proprietary tree ties which provide a cushion between the stake and the tree so there is no danger or rubbing. The last task is to mulch with bulky organic material to help soil moisture retention. Mulch over a radius of fifteen to eighteen inches to a depth of about two inches but keep it just clear of the tree stem to avoid any danger of collar rot.

Trees heeled in

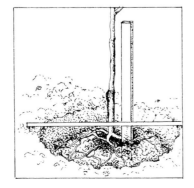

Planting young tree

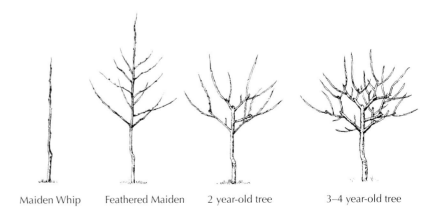

Maiden Whip   Feathered Maiden   2 year-old tree   3–4 year-old tree

PRUNING

It is beyond the intention of this book to go into great detail about pruning. Suffice to say that the restricted forms such as cordon, espalier, dwarf pyramid and fan are pruned in the summer. Its purpose is to inhibit their growth, necessary because they are grown in a confined way or in limited space, hence the term restricted. Summer pruning checks growth. The removal of leaves at the height of summer cuts down the supply of carbohydrates to the roots and therefore reduces the overall energy of the plant. It has other advantages too. It allows light and air on to the wood and on the fruit, helping to promote fruit buds for the next year and to improve the colour of the existing fruits. Winter pruning is applied to trees grown in the open, typically the bush, standard and spindlebush. Hard winter pruning has the effect of stimulating strong growth in the next growing season, hence is only usually applied to young trees whilst the framework is being formed. Established trees which are cropping well are pruned lightly.

It is not a wise policy to become hide-bound to one pruning system but apply whatever technique is necessary as circumstances dictate. There are occasions when summer pruning can be applied to trees normally pruned in the winter: the over vigorous, unfruitful bush or standard for example. Similarly, restricted forms can be winter pruned to restore a neglected cordon or espalier back to its original form or to reduce the length of over-long spur systems.

SPRING AND SUMMER TASKS

Feeding

It is necessary, on a regular basis, to replenish the main elements extracted from the ground by the tree for its growth and development and the creation of its crop. These are nitrogen (N), phosphorus (P) and potassium (K). In general terms nitrogen is necessary for growth, phosphorus (phosphates) for root development and fruit quality, and potassium (potash) for hardiness, fruit bud formation, fruit flavour and colour. There are also occasions when the minor elements, magnesium and calcium are needed. The trace elements are utilized in minute quantities and are therefore rarely needed except under certain special circumstances.

The most efficient way of supplying the trees' main nutrient requirements is to use the artificial fertilizers because they are concentrated, not bulky, and therefore easy to handle. They have their disadvantages, however, in that they do nothing for the soil structure and moisture retentiveness and very little for the soil fauna and flora, i.e. the life of the soil itself.

Nitrogen is usually applied in the form of sulphate of ammonia (21% N) on neutral to alkaline soils, and ammonium nitrate/lime (nitro-chalk) (21% N) on acid soils. Potassium is applied as sulphate of potash (48% K20), and phosphorus as superphosphates (18 – 19% P205) or triple phosphates (47% P205). Nitrogen and potash are applied annually and the phosphates triennially because the latter is used up much more slowly.

Apply the artificial fertilizers as a top dressing over the rooting area which is roughly equivalent to the spread of the tree and slightly beyond as follows: sulphate of potash in late January at three quarters of an ounce per square yard, superphosphates every third year in January at two ounces per square yard, triple superphosphates at three quarters of an ounce per square yard also in late January, sulphate of ammonia or ammonium nitrate/lime (nitro-chalk) at one ounce per square yard.

Dessert apples growing in grass and cookers are given twice this rate of nitrogen: in the case of dessert apples, to off-set the competition of grass for the nitrogen, and in the case of cookers to improve the size of the culinary fruits.

Turning now to the minor element magnesium. This is a constituent of the chlorophyll molecule necessary for the manufacture of sugars by the leaf. The trouble is, it is easily leached out of the leaves by rain therefore in wet summers and in areas of high rainfall expect a magnesium deficiency and act beforehand. Apply magnesium sulphate at twice the rate advocated

for potassium, i.e. one and a half ounces per square yard in early April. If, despite this, magnesium deficienty manifests itself, and it will in very wet summers, supplement by applying two or three foliar applications of magnesium sulphate (Epsom Salts) at fortnightly intervals. This is usually throughout the months of June and early July. Artificial fertilizers supply the necessary nutrients efficiently, but bulky organics also have their place in the feeding programme. Whilst they are not high in nutrient content compared with artificials, they are invaluable in other important respects. They maintain the life of the soil – the soil's fauna and flora – which in turn create and maintain the soil structure and its retentiveness. They also break down the various organic materials into a form which can be assimilated by the roots.

Closely spaced plants which are competing with each other, such as the restricted forms, will certainly benefit from a mulch of well rotted manure, compost, compost or mushroom compost. Spread along each side of the row or around the plant to a depth of about two or three inches but always keep it just clear of the stem of the tree so there is no danger of rotting diseases being encouraged. The mulch should be topped up annually in the spring.

The young tree in the orchard should also be mulched for the first four years. Thereafter the orchard can be grassed down so long as a clean area of about eighteen inches radius is maintained around the trunk of each tree. Remember, grass competes with the tree for nitrogen and extra nitrogen will be necessary, certainly for the cooking apples to give a good size.

Watering

Water in times of drought, especially the newly planted and young trees not yet fully established. Water is essential so that the trees grow well and form a strong framework. Just as importantly, irrigate if possible established trees carrying a heavy crop. Trees under stress due to lack of moisture are liable to suffer excessive fruit drop and go into a biennial pattern of bearing, producing a crop every other year. The relevant period for irrigation is the summer months of June, July and August. Apply about two inches (nine gallons per square yard) of water over the rooting area about once every seven to ten days starting in June and finishing in late August. Obviously stop when the rain restores the balance. To lessen the risk of fungal disease, apply the water over the ground rather than the foliage.

Thinning

Thinning is necessary in the event of a heavy fruit set to avoid too many small fruits and to lessen the risk of the onset of biennial bearing to which certain cultivars are prone. Remember, there will be a natural shedding of fruitlets and this takes place in late June and the first half of July. It is called the June drop. The main thinning should be done after this event but a little can

be carried out before by removing any malformed and diseased fruits. Thin dessert apples to about four to six inches apart, and culinary fruits, where a larger size is wanted, to six to nine inches. Obviously, remove the worst and leave the best. The King fruit, i.e. the one in the centre of the truss, is usually the largest and can be left, but check that it is not malformed at the stalk end. Weak trees should be thinned more drastically than the strong ones. Vigorous trees with a good show of leaves can be allowed to carry more fruit than the guidelines just given.

The Time to Pick

When the fruit is ready it leaves the spur easily. Lift the apple in the palm of the hand and give it a slight twist when it should part without undue effort. Avoid any finger pressure as this will bruise and spoil the appearance of the fruits. For the same reason, at all times treat the apples gently. Bruised fruits will not keep.

Other indications that the harvest is nigh are that the fruits take on brighter colours, and the pips turn from white to brown. Windfalls on the ground are another indication that they are ready, gales aside of course. Do not pick all the fruit at once, but practise colour picking, which means picking over a number of times, starting with those exposed to the most sun, leaving the inside fruits to the last so that they may eventually colour up.

Storage

Early apples do not keep, they should be eaten straight off the tree or within a few days of picking. July and August apples come within this category. The essential conditions required for good keeping are coolness, darkness, high humidity and ventilation. Not too much ventilation or the fruits will shrivel, nor too little or they will be destroyed by lack of oxygen. Do not keep mid-season apples with the lates because the volatile substances (that lovely smell they give off) will hasten the ripening of the lates. Store each cultivar separately as much as possible.

The fruit store should be able to maintain a cool, even temperature, ideally at about 37° F (2.7° C). Commercially this is done by refrigeration. On an amateur basis, the best one can usually achieve is about 42° to 47°F (5° to 7°C). A well built garden shed situated in the shade is fine. A garage comes second best, so long as frost does not penetrate the store and it is secure against mice. Do not keep apples in the loft as quite high temperatures can build up in this area when the sun is shining.

The construction of the container in which the apples are kept is just as important. It must allow air movement through the fruits and over the top. Wooden apple boxes with slatted sides and base and corner posts are ideal and so are wooden Dutch tomato trays. Store only sound fruit and inspect them regularly to remove rots. Remember all apple cultivars have their season: an

October apple is at its best in this month; before then it is immature and after, it is over the top. A cultivar with a season January to April, Sturmer Pippin for example, is ready to eat within this period and not before. Another very good way of keeping apples is in a polythene bag. This material maintains high humidity and so prevents the fruit from shrivelling too quickly. However, the apple must be allowed to breathe. The skin of the bag should be perforated with a hole the diameter of a pencil for every pound of fruit, and the top of the bag folded over rather than sealed. Use clear polythene so that the apples can be seen and any rots removed if necessary. The required conditions of coolness, darkness and ventilation still apply.

Pests and Disease Control

Unfortunately apples can be effected by a vast range of pests and diseases, not forgetting our feathered friends, the Bullfinch in the winter, and the Blackbirds, Thrushes and Starlings in the summer. Space permits mention only of the most important. Incidentally, never spray insecticides at blossom time because of the risk to bees and other pollinating insects. To spray or not to spray, that is the question. For those gardeners that want perfect fruits, destruction of the pathogens by chemical means is the usual answer at the present time. Scientists are aware of the public's growing reaction to the use of such chemicals and are working towards non-chemical means of control: the use of predators and parasites are typical examples. For the non-perfectionist, indeed for any good gardener, the other way of dealing with the problem is to practise garden hygene. Keep things tidy and clean. Keep the trees properly fed and pruned. Remove or burn the prunings, rots and other debris which might harbour trouble. Having said this, unsprayed trees will not yield perfect fruits and the gardener will have to accept that this is in the natural order of things.

Insects In order of their appearance. It is in spring that the over-wintering eggs of aphids and caterpillars hatch out. Once hatched they migrate to the opening flower and leaf buds.

Aphids are extremely debilitating creatures because they suck the sap out of the plant. They have the capability of multiplying rapidly and can very quickly infest the whole tree. Signs of their attack are severe leafcurling, stunted and distorted growth and dwarfed malformed fruits.

Woolly Aphid is another aphid, very different in its appearance and mode of attack. It does not over-winter in egg form but as a complete insect. It is rather an immobile aphid which usually lives in the crevices of the older wood, in wounds, cracks and under the bark. The Woolly Aphid is covered with a woolly-like substance, presumably as a form of camouflage, hence the name. Other indications of its attack are large, gall-like swellings on the wood where it lives.

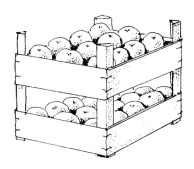

Winter Moth Caterpillars are looper caterpillars which eat the spring foliage and the developing flowers. Damage can be serious.

Apple Sawfly the larvae – white maggots – either tunnel into the fruitlets which then drop, or graze a line along the surface, usually around the stalk end, causing a ribbon-like scar.

Fruit Tree Red Spider Mite a sap-sucking insect which lives on the underside of the leaves and which thrives in a hot summer. It is very small and a hand lens is the best way of seeing it. Red Spider Mite is a very difficult insect to control by chemical means. It is best left to natural predators

Codling Moth the most important of the insect pests. The larvae is a caterpillar which eats into the centre of the apple causing a most unsavoury mess. The moth lays her eggs in mid to late June. The use of pheromone traps, which attract the male moths, reduce the number of fertilized females thus resulting in the reduction of egg laying.

Diseases The most important are scab, mildew and canker. There are a few cultivars which have some degree of resistance to one or other of these diseases and this is mentioned in the cultivar descriptions.

Apple Scab a fungus which causes black scab-like lesions on the skin of the fruit and, in severe cases, cracking and malformation. It also attacks the leaves and sometimes the wood. On the leaves it manifests itself as brown or olive green spots resulting in early leaf fall. On wood it shows as small blister-like pimples on the young shoots and can later lead to the wood-rotting disease apple canker. Scab is worst in a wet summer and in areas of high rainfall.

Apple Mildew – Powdery Mildew a fungal disease mainly of the apple foliage, though it can dull and russet the skin of the fruit. It shows as a powdery or mealy coating of the leaves and shoots as they move into growth in the spring. Mealy rosettes of leaves with little or no growth is another sign. If left uncontrolled, mildew will persist and multiply throughout the tree during the growing season and is a most debilitating problem. Cut off and burn every infected shoot as they become obvious in the early spring. Remember there are cultivars available that are resistant or partly resistant.

Apple Canker another serious disease of apples to which certain cultivars are prone, for example James Grieve, Spartan and Cox's Orange Pippin. It is more prevalent on trees growing in badly drained land and in wet areas. Canker is a wood rotting disease showing first as sunken oyster-shaped areas, which if left uncontrolled will eventually girdle and kill the infected branch and obviously, if on the main trunk, can result in the death of the tree. Cut out the diseased wood and paint the wound with a fungicidal tree paint. Girdled, dead or dying branches must be removed and burnt. An untreated canker is a reservoir of infection.

Bitter Pit is a form of calcium deficiency within the fruit exaccerbated by the excessive use of nitrogenous fertilizer, over hard winter pruning, and lack of water particularly at the critical stage of fruit development. Some cultivars are very prone: for example Newton Wonder, Bramley's Seedling and Egremont Russet. It produces slightly sunken pits on the surface of the skin and small brown areas in the flesh immediately beneath the pits and scattered throughout the apple. It is thought that the onset of this disease is caused by the lack of water at a critical time and a calcium deficiency. It can be mitigated by mulching to retain soil moisture and also watering in times of drought.

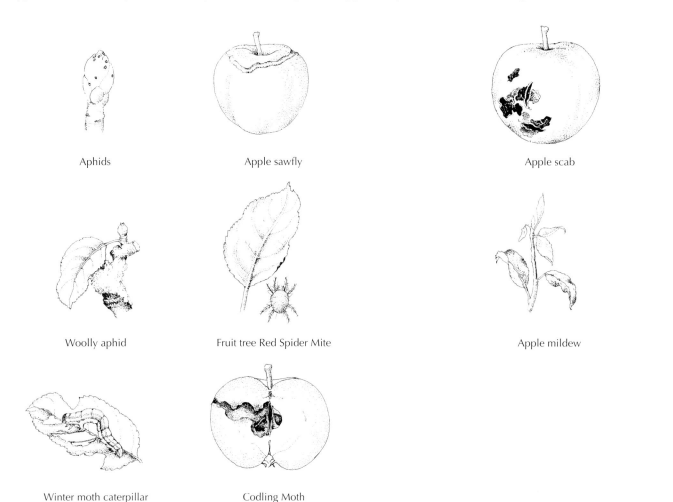

Aphids

Apple sawfly

Apple scab

Apple canker

Woolly aphid

Fruit tree Red Spider Mite

Apple mildew

Bitter pit

Winter moth caterpillar

Codling Moth

RECOMMENDED CULTIVARS

The list below indicates the apple cultivars, both dessert and culinary, which are generally considered to be of the highest quality. In most cases, this has been established through reputation over a great number of years but it also includes some more recent and promising introductions.

Dessert Apples

Adam's Pearmain
Alkmene (Red Windsor)
Ashmead's Kernel
Blenheim Orange
Cornish Gilliflower
Cox's Orange Pippin
D'Arcy Spice
Delcorf
Discovery
Egremont Russet
Fiesta
Gala
George Cave
Greensleeves

Herefordshire Russet
Holstein
Irish Peach
James Grieve
Jonagold
Jupiter
Kidd's Orange Red
Laxton's Epicure
Laxton's Fortune
Lord Hindlip
Lord Lambourne
Merton Charm
Mother
Orleans Reinette

Pitmaston Pineapple
Pixie
Red Falstaff
Ribston Pippin
Ross Nonpareil
Scrumptious
St Edmund's Pippin
Spartan
Sunset
Suntan
Tidicombe Seedling
William Crump
Winter Gem

Culinary Apples

Annie Elizabeth
Blenheim Orange
Bountiful
Bramley's Seedling
Don's Delight
Encore
George Neal
Golden Noble
Monarch
Newton Wonder
Norfolk Beauty
Rev. W Wilks
Warner's King

POLLINATION TABLE

T = Triploid   B = Biennial or irregular in flowering   S/f = Self fertile

GROUP 1
Gravenstein (T)
Stark's Earliest

GROUP 2
Adam's Pearmain
Alkmene (Early Windsor)
Beauty of Bath
Bismarck (B)
Devonshire Quarrenden (B)
Egremont Russet
George Cave
Idared
Irish Peach
Keswick Codlin (B)
Laxston's Early Crimson
Lord Lambourne
Margil
McIntosh Red
Merton Charm
Norfolk Beauty
Owen Thomas
Red Falstaff (S/f)
Rev. W. Wilks (B)
Ribston Pippin (T)
Ross Nonpareil
St. Edmund's Pippin
Tidicombe Seedling
Warner's King

GROUP 3
Allington Pippin (B)
Arthur Turner
Belle de Boskoop (T)
Belle de Pontoise (B)
Blenheim Orange (T) (B)
Bountiful
Bramley's Seedling (T)
Brownlees Russet
Catshead
Charles Ross
Cortland
Cox's Orange Pippin
Crispin (T) (B)
Delcorf
Discovery
Don's Delight
Duchess's Favourite
Emneth Early (B)
Emperor Alexander
Fiesta
Granny Smith
Greensleeves
Grenadier
Hambledon Deux Ans
Herefordshire Russet
Holstein (T)
James Grieve
John Standish
Jonathan
Jupiter (T)
Katy
Kidd's Orange Red

Lane's Prince Albert
Laxton's Epicure
Laxton's Fortune (B)
Lord Grosvenor
Lord Hindlip
Malling Kent
Mère de Ménage
Merton Knave
Merton Worcester
Miller's Seedling (B)
Nonpareil
Peasgood Nonsuch
Pitmaston Pineapple
Queen
Red Delicious
Red Devil (S/f)
Rival (B)
Rosemary Russet
Saturn (S/f)
Scotch Bridget
Scrumptious (S/f)
Spartan
Stirling Castle
Sturmer Pippin
Sunset
Tidicombe Seedling
Tom Putt
Tydeman's Early Worcester
Upton Pyne (S/f)
Wagener (B)
Wealthy
Winter Gem
Worcester Pearmain

GROUP 4
Annie Elizabeth
Ashmead's Kernel
Autumn Pearmain
Barnack Beauty
Braeburn
Chiver's Delight
Claygate Pearmain
Cornish Aromatic
Cornish Gilliflower
Cox's Pomona
D'Arcy Spice
Duke of Devonshire
Dumelow's Seedling
Ellison's Orange
Encore
Gala
Gladstone (B)
Golden Delicious
Golden Noble
Harvey
Herring's Pippin
Hoary Morning
Howgate Wonder
Ingrid Marie
Jester
Jonagold (T)
King's Acre Pippin
Lady Sudeley
Lady Henniker
Laxton's Superb (B)
Limelight
Lord Burghley

Lord Derby
Monarch (B)
Orleans Reinette
Pixie
Sweet Society
Tower of Glamis
Tydeman's Late Orange
Winston

GROUP 5
Gascoyne's Scarlet (T)
King of the Pippins (B)
Merton Beauty
Mother
Newton Wonder
Norfolk Royal
Royal Jubilee
Suntan (T)
William Crump

GROUP 6
Bess Pool
Court Pendu Plat
Edward VII

GROUP 7
Crawley Beauty

CLASSIFICATION TABLE

KEY
Numbers indicate months in which the apple should be eaten

c = Culinary   † = No ribs or slight trace   ‡ = Indefinitely ribbed
d = Dual purpose   # = Fairly prominently ribbed   # = Boldly ribbed

FLAT ⬭	ROUND ◯	CONICAL ⬯	OBLONG AND OVAL ⬭ ⬭
GROUP 1 GREEN – SMOOTH SKINNED – ACID			
Stirling Castle 9/12 † c	Grenadier 8/10 # c	Emneth Early 7/8 # c	
Bramley 11/3 # c		Lord Grosvenor 8/10 # c	
	Warner's King 9/2 # c		Delcorf 9/10 ‡
	Edward V11 12/4 † c	Lord Derby 10/12 # c	Catshead 10/1 # c
	Tower of Glamis 11/2 # c		
GROUP 2 GREEN – SMOOTH SKINNED – SWEET			
	Limelight 9/11 #		
	Granny Smith 1/4 ‡		Greensleeves 9/11 †
GROUP 3 WHITISH CREAM or YELLOW SKIN – SWEET or ACID			
	Norfolk Beauty 10/12 # c	Keswick Codlin 9/10 # c	Greensleeves 9/11 †
	Golden Noble 10/12 † c	Rev W Wilks 9/10 ‡ c	Royal Jubilee 10/12 # c
		Harvey 9/1 ‡ c	Golden Delicious 11/2 ‡
		Arthur Turner 9/11 ‡ c	Crispin 12/2 # d
GROUP 4 GREEN or YELLOW SKIN – FLUSHED and/or STRIPED – ACID			
	George Neal 8/10 ‡ c	George Cave 8 †	
	Tom Putt 9/11 # c	Don's Delight 9/2 # c	
Queen 9/12 ‡ c	Hoary Morning 10/1 † d	Howgate Wonder 10/3 ‡ c	
Belle de Pontoise 11/3 #	Monarch 11/1 † c	Scotch Bridget 10/12 # c	
	Newton Wonder 11/3 † c	Upton Pyne 11/2 # d	
Bramley 11/3 # c	Dumelow's Seedling 11/3 † c		
	Lane's Prince Albert 12/3 ‡ c		Encore 12/4 ‡ c
	Crawley Beauty 12/3 † c		Annie Elizabeth 12/6 † c
GROUP 5 GREEN or YELLOW SKIN – FLUSHED and/or STRIPED and SUB-ACID			
Owen Thomas 8/9 #	Irish Peach 8/9 ‡	Lady Sudeley 8/9 ‡	
	Beauty of Bath 8 †	Miller's Seedling 8/9 ‡	
	Laxton's Fortune 9/10 ‡	Laxton's Epicure 8/9 †	
	Merton Charm 9/10 †	James Grieve 9/10 ‡	
	Bountiful 9/11 ‡ c	Emperor Alexander 9/11 † c	Gravenstein 9/12 # d
	Lord Lambourne 9/11 †		
	Peasgood Nonsuch 9/12 † c	Alkmene (Early windsor) 10/11 †	
	Rival 10/12 ‡	Cox's Pomona 10/12 # c	Jester 10/11 #
	Cortland 10/2 †	Charles Ross 10/12 †	Winter Gem 10/3 #
	Chiver's Delight 11/1 ‡	Sturmer Pippin 1/4 ‡	
Wagener 12/4 #		Jonagold 11/2 †	Braeburn 1/3 †

GROUP 6 OVER HALF FLUSHED BRIGHT RED			
Discovery 8/9 †	Stark's Earliest 7/8 †	Laxton's Early Crimson 7/8 †	
Duchess's Favourite 8/9 †		Gladstone 7/8 #	
Devonshire Quarrenden 8/9	Tydeman's Early Worcester 8/9 ‡	Katy 9/10 †	
		Worcester Pearmain 9/10 †	
	Scrumptious 9 #		
	Merton Knave 9 † Merton Worcester 9/10 † Herring's Pippin 9/11 # d		
	Wealthy 9/12 ‡	Norfolk Royal 9/12 ‡	
	Merton Beauty 9/10 †	Saturn 9/2 #	
	Red Devil 10/11 # Sweet Society 10/1 †	Gala 10/1 ‡	
Fiesta 10/1 † McIntosh Red 10/12 ‡ Spartan 10/2 ‡ Mother 10/11 ‡			
Ingrid Marie 10/12 † Gascoyne's Scarlet 10/1 ‡		Jupiter 11/1 ‡	Red Falstaff 10/12 †
Mere de Menage 11/2 #		Bismarck 11/2 #	Jonathan 11/2 #
Idared 11/4 ‡		Malling Kent 11/2 ‡	Red Delicious 12/3 #
	John Standish 12/2 †	William Crump 12/2 ‡	
GROUP 7 REINETTES			
	Ellison's Orange 9/10 †	Autumn Pearmain 9/11 ‡	
	Tidicombe Seedling 9/12 †		
	Ribston Pippin 10/12 #	Allington Pippin 10/12 †	King of the Pippins 10/12 †
	Sunset 10/12 † Ross Nonpareil 10/12 †		
Suntan 11/1 †		Margil 10/1 #	
Blenheim Orange 11/1 †		Cox's Orange Pippin 10/1 †	
Orlean's Reinette 11/1 †		Kidd's Orange Red 11/1 ‡	Lady Henniker 11/1 #
		Laxton's Superb 11/1 ‡	
		Holstein 11/1 †	
		Rosemary Russet 11/3 ‡	Cornish Gilliflower 11/3 #
		Adam's Pearmain 11/3 †	
		Hambledon Deux Ans 11/4 ‡	
	Bess Pool 12/2 †		Claygate Pearmain 12/2 ‡
Pixie 12/3 †		Cornish Aromatic 12/3 #	Barnack Beauty (oval) 12/3 †
		Lord Hindlip 12/3 #	
		King's Acre Pippin 12/3 †	
		Belle de Boskoop 12/4 ‡	
Court Pendu Plat 12/4 †		Tydeman's Late Orange 12/4 †	
	Lord Burghley 1/4 ‡	Winston 12/4 †	
GROUP 8 RUSSETS			
	St Edmund's Pippin 9/10 †	Pitmaston Pineapple 9/12 †	D'Arcy Spice 12/4 #
Egremont Russet 10/12 †		Herefordshire Russet 10/1 †	
Ashmead's Kernel 12/2 ‡			
Brownlees' Russet 12/3 ‡			
	Nonpareil 12/3 †		
	Duke of Devonshire 1/3 †		

INDEX OF APPLES

ACKNOWLEDGEMENTS

My warmest thanks to the following people for the help I have received with the new edition of this book:
Kevin Croucher of Thornhayes Nursery, Sally Saunders of Hafod Farm, Martin Crawford, Richard Smedley of Four Elms Fruit Farm, RHS Wisley, Simon Maughan and Diana Gilding at RHS Rosemoor, Michael Honnor, Peter Prew and most of all my sincere thanks to Harry Baker who has been my guide and mentor throughout.

BIBLIOGRAPHY

Baker.H. *The Fruit Garden Displayed,* 1986, Royal Horticultural Society, Cassell.

Beach, S.A. *The Apples of New York Vol 1 & 2,* 1905, Albany, New York.

Bultitude John. *Apples,* 1983, Macmillan Reference Books.

Bunyard Edward A. *A Handbook of Hardy Fruits, Apples and Pears.* 1920. London.

Hills Lawrence D. *The Good Fruit Guide,* 1984, Henry Doubleday Research Association.

Hogg Robert. *The Fruit Manual, 5th Edition,* 1884 London.

Maund B. *The Fruitist,* 1845–51, Groombridge & Sons, London.

Morgan J. *The Book of Apples,* 1993, Ebury Press, London.

Roach F A. *Cultivated Fruits of Britain,* 1985 Basil Blackwell. U.S.A., Oxford.

Scott J. *The Orchadist, c* 1873, London.

Simmons A.F. *Simmons' Manual of Fruit,* 1978, David & Charles.

Smith Muriel W G. *National Apple Register of the United Kingdom* 1971. Ministry of Agriculture, Fisheries & Food, London.

Taylor H V. *The Apples of England,* 1948, Crosby, Lockwood, London.

Gardeners Chronical, 1883, 1884, 1886, 1900, 1908, 1930.

The Herefordshire Pomona, 1878–85.

Transactions, Journals & Proceedings of the Royal Horticultural Society, 1805 onwards.